REVIEWS IN ECONOMIC GEOLOGY

(ISSN 0741-0123) Volume 3

EXPLORATION GEOCHEMISTRY: DESIGN AND INTERPRETATION OF SOIL SURVEYS

in cooperation with
The Association of Exploration Geochemists

ISBN 0-9613074-2-0

The Authors:

W. K. Fletcher
Department of Geological Sciences
6339 Stores Road
University of British Columbia
Vancouver, BC   Canada, V6T 2B4.

S. J. Hoffman
Selco Division
BP Resources Canada Limited
700-890 West Pender Street
Vancouver, BC   Canada, V6C 1K5.

M. B. Mehrtens
U.S. Minerals Exploration Company
141 Union Boulevard, Suite 100
Lakewood, Colorado 80228

A. J. Sinclair
Department of Geological Sciences
6339 Stores Road
University of British Columbia
Vancouver, BC   Canada, V6T 2B4.

I. Thomson
Placer Development Limited
1055 Dunsmuir Street
Vancouver, BC   Canada, V7X 1P1

Series Editor: James M. Robertson
New Mexico Bureau of Mines & Mineral Resources
Campus Station
Socorro, NM 87801

SOCIETY OF ECONOMIC GEOLOGISTS

FOREWORD

Volume 3 of *Reviews in Economic Geology*—Exploration Geochemistry: Design and Interpretation of Soil Surveys—represents a major effort by and contribution from the Association of Exploration Geochemists (AEG) and especially its Vancouver connection. The volume draws extensively on the cumulative teaching, research, and industry experience of its five authors, and it contains numerous 'real-life' examples of exploration failures as well as successes. A preliminary version of this volume served as the text for a jointly sponsored Society of Economic Geologists (SEG)–AEG Short Course that was given in February, 1987, prior to the combined winter meeting of the SEG and annual meetings of the Society of Mining Engineers and A.I.M.E. in Denver, Colorado.

It has been a special pleasure to work with W. K. Fletcher (Department of Geological Sciences, U.B.C.) whose patience and sense of humor survived the herculean task of initial text and figure assembly. He met his deadlines despite the vagaries of changing figure specifications, colleagues' schedules, and the Canadian Postal System.

Volume 3 has benefited greatly from the professional attentions of Carol Hjellming (New Mexico Bureau of Mines and Mineral Resources editing staff) who now serves as the part-time official assistant to the Series Editor. In addition to performing more traditional editorial chores, Carol has been instrumental in setting up the procedures and print codes that allowed us to utilize the computer-driven typesetting equipment of the University of New Mexico Printing Plant.

Finally, I wish to acknowledge the continuing support, both moral and economic, of the New Mexico Bureau of Mines and Mineral Resources and its Director, Frank Kottlowski.

James M. Robertson
Series Editor
Socorro, N. M.
April, 1987

CONTENTS

FOREWORD ii
PREFACE.. v
BIOGRAPHIES vi

Chapter 1—GETTING IT RIGHT
INTRODUCTION 1
CHOICE OF METHODS 1
OPTIMIZING SURVEY TECHNIQUES 2
  BASIC OBJECTIVES 2
    Optimum Target Identification..................... 2
    Maximum Geochemical Contrast................... 3
    Minimum False Alarm Rate 5
    Cost Effectiveness................................ 6
  SURVEY PARAMETERS 7
  ORIENTATION STUDIES 10
    The Orientation Survey 10
    A Literature Study 10
    A Theoretical Orientation 10
SURVEY ORGANIZATION AND OPERATION 10
PROBLEM 1: MOLYBDENUM IN SIERRA LEONE..... 11
  OBJECTIVE.. 11
  DESCRIPTION OF THE AREA......................... 11
  ASSUMED 11
  QUESTIONS... 11
PROBLEM 2: EXPLORATION FOR BASE METALS IN
  A GLACIATED AREA OF CENTRAL NORWAY...... 12
  OBJECTIVE.. 12
  DESCRIPTION OF THE AREA......................... 12
  ASSUMED 13
  QUESTIONS... 14
REFERENCES 17

Chapter 2—GEOCHEMICAL EXPLORATION—THE SOIL SURVEY
INTRODUCTION 19
PHASE 1—THE OFFICE................................ 19
  THE UNCONFORMITY-RELATED URANIUM DEPOSIT....... 19
  THE EPITHERMAL GOLD DEPOSIT 21
PHASE 2—THE FIELD ORIENTATION VISIT 22
  THE UNCONFORMITY-RELATED URANIUM DEPOSIT 22
    Introduction 22
      Athabasca Basin............................... 23
      Thelon Basin.................................. 24
      Hornby Bay Basin............................. 24
      The Preliminary Field Visit—Is It Necessary? 25
  THE EPITHERMAL GOLD DEPOSIT 26
CONTINUED OFFICE PLANNING 26
  THE UNCONFORMITY-RELATED URANIUM DEPOSIT 26
  THE EPITHERMAL GOLD DEPOSIT 26
REGIONAL EXPLORATION 27
  THE UNCONFORMITY-RELATED URANIUM DEPOSIT 27
  THE EPITHERMAL GOLD DEPOSIT................. 28
THE ROUTINE SOIL SURVEY 28
  THE UNCONFORMITY-RELATED URANIUM DEPOSIT 28
  THE EPITHERMAL GOLD DEPOSIT 28

THE ULTIMATE TEST—THE DIAMOND DRILL
  PROGRAM.. 31
  THE UNCONFORMITY-RELATED URANIUM DEPOSIT 31
  THE EPITHERMAL GOLD DEPOSIT 33
CONCLUDING SUMMARY 33
ANSWERS TO EPITHERMAL GOLD DEPOSIT
  QUESTIONS 33
REFERENCES 38

Chapter 3—SOIL SAMPLING
INTRODUCTION 39
THE SOIL SURVEY AS PART OF THE EXPLORATION
  PROGRAM.. 40
AN EXPLORATION EXAMPLE—THE "QUICK AND
  DIRTY" VERSUS THE "SLOW AND
  PROFESSIONAL" APPROACH 41
GEOCHEMICAL FACTORS AFFECTING TRACE
  ELEMENT DISTRIBUTION: SOIL
  DESCRIPTIONS..................................... 43
  SAMPLE TYPE.. 45
  SAMPLE NUMBER 47
  TOPOGRAPHY.. 48
  SITE DRAINAGE AND GROUNDWATER SEEPAGE 48
  OVERBURDEN ORIGIN............................... 48
  SOIL pH 52
  TEXTURE 54
  SAMPLE DEPTH 54
  SOIL HORIZON 57
  ROCK TYPE.. 64
  CONTAMINATION 66
  COARSE FRAGMENTS................................ 66
  GAMMA COUNT AT SAMPLE SITE 70
  OTHER PARAMETERS—COMPOSITION AND/OR SITE 70
SUMMARY... 70
ACKNOWLEDGEMENTS 70
REFERENCES 70
APPENDIX I.. 71
APPENDIX II... 72
APPENDIX III.. 76

Chapter 4—ANALYSIS OF SOIL SAMPLES
INTRODUCTION 79
DISTRIBUTION OF TRACE METALS IN SOILS........ 80
SAMPLE PREPARATION................................ 81
SAMPLE DECOMPOSITION 82
  INTRODUCTION 82
  STRONG DECOMPOSITIONS 82
  PARTIAL EXTRACTIONS 84
ANALYTICAL METHOD................................. 85
QUALITY CONTROL AND RELIABILITY.............. 87
  RANDOM ERRORS AND PRECISION 87
  SYSTEMATIC ERRORS 89
  CONTAMINATION 90
  DRIFT 91
  INTERFERENCES 91

Monitoring Systematic Errors........................ 91	Outcrop.. 126
Submission of Samples for Analysis............ 92	Overburden.. 126
An Analytical Case History: Tin Exploration in Eastern North America... 93	pH-Eh/Element Mobility........................... 126
	Anomaly Followup.................................... 127
References.. 96	References.. 128

Chapter 5—STATISTICAL INTERPRETATION OF SOIL GEOCHEMICAL DATA

Chapter 7—CASE HISTORY AND PROBLEM 1: THE TONKIN SPRINGS GOLD MINING DISTRICT, NEVADA, U.S.A.. 129

Introduction.. 97	
Basic Statistics.. 99	
General Statement.................................... 99	**Chapter 8—CASE HISTORY AND PROBLEM 2: COED–Y–BRENIN PORPHYRY COPPER, NORTH WALES, GREAT BRITAIN**............... 135
Central Tendency..................................... 99	
Arithmetic Mean.................................... 99	
Median... 99	
Mode.. 99	**Chapter 9—CASE HISTORY AND PROBLEM 3: THE VOLCANOGENIC MASSIVE-SULFIDE TARGET**
Geometric Mean.................................... 99	
Dispersion.. 99	Preliminary Studies.................................. 139
Range... 99	Field Orientation...................................... 139
Variance... 99	Continued Office Planning....................... 141
Standard Deviation............................... 99	Property Evaluation.................................. 141
Percentiles.. 100	Anomaly Followup—Drill Testing............. 141
Histograms.. 100	Answers... 141
Continuous Distributions........................ 100	Summary... 146
Standard Normal Distribution................ 101	References.. 146
Lognormal Distributions......................... 101	
Fitting a Normal Curve to a Histogram.......... 102	**Chapter 10—CASE HISTORY AND PROBLEM 4: THE VOLCANOGENIC MASSIVE SULFIDE, A SECOND EXAMPLE**
Cumulative Distributions........................ 103	
Confidence Limits.................................... 103	
F and t Tests.. 103	Preliminary Studies.................................. 147
Probability Graphs.................................. 104	Field Observations.................................... 147
Correlation... 106	Continued Office Planning....................... 147
Introduction... 106	Property Evaluation.................................. 147
Analysis of a Matrix of Correlation Coefficients... 107	Anomaly Followup—Drill Testing............. 152
	Answers... 152
"Correlation" of Populations................... 109	Summary... 154
Correlations Among Percentage Data........... 109	
Autocorrelation.. 109	**Chapter 11—CASE HISTORY AND PROBLEM 5: A COPPER PROPERTY**
Possible Problems in Correlation Studies......... 110	
Simple Linear Regression.......................... 110	Preliminary Studies.................................. 155
Introduction... 110	Field Orientation...................................... 155
Summary of Formulae............................. 111	North Cirque.. 155
Some Applications of Linear Regression.......... 111	North Creek... 155
Degree of Fit.. 112	North Tip... 155
Errors in Both Variables.......................... 113	Tabletop Highlands................................ 155
Chi Square Distribution.............................. 113	South Cirque.. 155
Introduction... 113	Main Valley.. 159
Goodness of Fit.. 113	L. Mountain.. 159
Two-way Contingency Tables................. 114	Pegmatite Hill... 159
Final Remarks... 114	Soils.. 159
References.. 115	Semiregional Exploration........................... 159
	Continued Office Planning....................... 159
Chapter 6—MODELS, INTERPRETATION, AND FOLLOWUP	Property Evaluation.................................. 159
	Geology.. 160
Models.. 117	Geophysical Surveys............................... 160
General Background................................ 117	Soil Geochemistry.................................. 165
Landscape Geochemistry........................ 117	Anomaly Followup—Drill Testing............. 170
Idealized Models..................................... 118	Answers... 170
Examples.. 119	Summary... 180
Applications... 121	References.. 180
Interpretation... 122	
Landscape/Topography........................... 126	TABLES OF CONVERSION FACTORS. **Inside back cover**

iv

PREFACE

The principles and practical considerations underlying utilization of soils as a medium for exploration geochemistry are well described in several textbooks. Moreover, not only are soil surveys routinely undertaken in such diverse environments as tropical rainforests and arctic permafrost, soils are probably the most frequently collected and analyzed medium in exploration geochemistry. What, then, is the justification for devoting the third volume in the Society of Economic Geologists *Reviews in Economic Geology* to this apparently routine, well established prospecting method?

Unfortunately, it is the experience of the contributors to this volume that effectiveness of soil surveys is often compromised when the conceptual simplicity of the method leads to its unthinking application. For example, failure to appreciate the characteristics of the geochemical environments of a landscape can lead to collection of the wrong sample material or choice of unsuitable methods of sample preparation and analysis. Similarly, emphasis on speed rather than quality of sampling, rigid adherence to standard laboratory methods, and simplistic interpretations of high values can result in exploration dollars being wasted on false anomalies while genuine, but more subtle, anomalies go unrecognized or are assigned low priorities. In contrast to the foregoing, rational application of soil surveys depends on the successful selection and linkage of appropriate methods of sample collection, analysis, and interpretation—often on the basis of an initial orientation survey. Decisions must be made at each step and an error at any single step may jeopardize the entire exploration effort.

Chapters in this volume discuss each step in the soil survey from sample collection through analysis and statistical interpretation of the data to selection of targets for followup. Factors to be considered and the decisions that must be made are illustrated by numerous examples and case histories. However, rather than presenting the case histories in a simple narrative fashion, we have attempted to challenge the reader, by asking questions as each case history unfolds, to become a participant in the exploration process. In some—but not all—cases we have provided answers (or our opinions as to what reasonable answers might be). The case histories are largely from our own experience and many reflect our geographical bias towards northern glaciated regions. We do not believe this to be a deficiency insofar as this volume is intended not as a comprehensive guidebook to interpretation of soil surveys but as an introduction to undertaking surveys in a thoughtful and logical fashion. Indeed this volume will be a success if its omissions provoke you into asking similar (though not necessarily the same) questions of your own geochemical landscapes and soil surveys.

ACKNOWLEDGEMENTS—In preparing this volume the authors were assisted by many individuals and organizations who generously contributed their time, technical facilities, experience, and comments. We are especially grateful to our respective employers for their support and freedom to use company case histories even when these were not entirely flattering. Of the many who encouraged and assisted, the following deserve special mention: Riofinex and CARGO Partners for giving M. B. Mehrtens use of the Coed-y-Brenin and Tonkin Springs case histories, respectively; BP–Selco, D. K. Mustard, C. M. H. Jennings and G. G. Mitchell for their assistance to Stan Hoffman; and Placer Development Limited for their support of Ian Thomson's contributions. Donna M. Baylis greatly assisted K. Fletcher in editing and preparing the text. Finally, we must acknowledge the patience of Jamie Robertson, Series Editor, *Reviews in Economic Geology*, and the support of both the Association of Exploration Geochemists and the Society of Economic Geologists.

W. K. Fletcher
Chairman
Short Course Committee
Association of Exploration Geochemists

BIOGRAPHIES

W. K. FLETCHER received both his B.Sc. in Geology and Ph.D. in Applied Geochemistry at the Imperial College of Science and Technology, University of London. He joined the Department of Geological Sciences, University of British Columbia in 1968 and is now an Associate Professor. During leaves of absence he has been Chief Geochemist to MIN-DECO (Zambia) and Geochemist and Team Leader to the United Nations Project at the Southeast Asia Tin Research and Development Centre in Malaysia. He is an author of more than fifty scientific papers on applied geochemistry and of a textbook *Analytical Methods in Geochemistry Prospecting*. He is a former council member of the Association of Exploration Geochemists and is the current Chairman of the association's Short Course Committee.

STAN J. HOFFMAN received his B.Sc. in Geology and Chemistry from McGill University and his M.Sc. and Ph.D. in exploration geochemistry at the University of British Columbia. He has more than twenty years of field-related experience working for a number of mineral exploration companies including INCO, Amax, Rio Tinto and BP Minerals. He is currently Senior Geochemist for the Selco Division of BP Resources Canada Limited and is based in Vancouver. He has actively advanced use of geochemistry by the mineral exploration community through short courses, organization of symposia (GOLD-81 and GEOEXPO/86) and compilation of a manual on "Writing Geochemical Reports" (Association of Exploration Geochemists, Special Volume 12). He is a Member of Council of the Association of Exploration Geochemists and President of the AEG for 1987–1988.

MIKE B. MEHRTENS has a B.Sc. in Geology and received his Ph.D. in Applied Geochemistry (1966) from the Imperial College of Science and Technology, University of London. He has been employed in mining and exploration for base and precious metals in Zambia, South Africa, the United Kingdom, Saudi Arabia, and Panama. He is currently President of U.S. Minerals Exploration Company (USMX) based in Denver, Colorado.

ALASTAIR J. SINCLAIR, P. Eng., has had 24 years of teaching, research, and consultancies in the mineral industry. He has taught economic geology, geological data analysis, and geostatistics during a 22-year career in the Department of Geological Sciences, The University of British Columbia, where he is now Professor and Head of Department. During that period he has been involved in a broad range of local and international consulting for numerous mining companies as well as the Provincial government and the United Nations, particularly in the fields of mineral property evaluation, mineral exploration data analysis, and geostatistical ore reserve estimation. His research and field work since the late 1950's has led to more than 100 scientific and technical publications, many of which have direct application to exploration for and evaluation of mineral deposits. Dr. Sinclair has served in executive capacities for a variety of professional organizations including The Canadian Institute of Mining and Metallurgy, The Mineral Deposits Division of the Geological Association of Canada, and the Society of Economic Geologists and is active in the Association of Professional Engineers of British Columbia.

IAN THOMSON received his Ph.D. in Applied Geochemistry for research at Imperial College of Science and Technology, University of London. On graduation he joined Barringer Research Limited where, over an eight-year period, he was involved in the development of field and analytical techniques in geochemistry, and consulting and project work in Canada, the U.S.A., Central and South America, the southwest Pacific islands and the Middle East. In 1978 he joined the Ontario Geological Survey where, for three years, he was involved in a number of studies including deep-overburden-till sampling in the Abitibi Clay Belt and an examination of the impact of acid rain on the geochemistry of lakes. He joined Placer Development Limited in 1981 as Senior Geochemist and is now Manager Western Canada Exploration and based in Vancouver. He is a former President of the Association of Exploration Geochemists and author of numerous publications on applied geochemistry.

Chapter 1

GETTING IT RIGHT

I. Thomson

INTRODUCTION

As mineral exploration becomes increasingly difficult, costly and competitive, success is essential; there is no room for waste or inefficiency. Exploration must be truly cost effective. The present book is concerned ultimately with the interpretation of geochemical surveys. However the data to be interpreted are the product of the field survey and thus only as good as the work that went into these earlier phases. The truism "garbage in—garbage out" is as relevant here as anywhere.

Every exploration geochemical survey has three component parts:

(1) Sampling
(2) Analysis
(3) Interpretation

These are independent yet interdependent functions. Failure to execute one step correctly will negate all efforts in the succeeding steps. By and large the function that is most costly, and certainly most difficult to repeat, is the field survey. Any deficiences at this stage will have fatal effects on the remainder of the project. Analysis of the samples is, indeed, costly and an area of necessary concern. However, if samples have been collected properly it is not unreasonable to suppose that they can be reanalyzed should this be deemed necessary or useful. Ultimately interpretation, provided that sampling and analysis are reliable, is an exercise that can be repeated many times using a variety of techniques or models depending on supplementary information available and the skills and prejudices of various geologists or geochemists.

The design and execution of a geochemical survey is thus crucial to its success. Surveys can be, and are, optimized to find specific targets in particular environments. Such fine tuning requires an understanding of applied geochemistry, knowledge of the environment in which the survey will be carried out and an appreciation of the target being sought.

Before considering these points in more detail, it is worth defining the nature of a geochemical survey in more general terms and establishing clearly the role of the survey in an exploration program.

The basic premise of exploration geochemistry is that the systematic sampling and analysis of naturally occurring materials will reveal features indicative of the presence of potentially economic mineralization. This is a deceptively simple statement for it begs the questions—what materials should be sampled?; how and for what entities should these samples be analyzed?; and what features will be revealed? We will consider these points in more detail in later sections.

The key wording here is "systematic sampling and analysis." Regular and consistent application of a technique across a property should produce a common database, a synoptic picture of the distribution of elements or compounds, that will meet the twin objectives of any exploration survey:

(1) Identification of targets likely to represent potentially significant mineralization.
(2) Confident elimination of barren ground.

It is thus important to know both the advantages and the limitations of any survey or survey technique. A good geochemist should be able to interpret every feature on a geochemical map: the lows as well as the highs. This is not always the case. A frightening number of dollars have been spent on conventional soil surveys over areas of deep transported overburden. The resulting geochemical patterns are typically flat and uninteresting. Many interpretations have concluded, wrongly, that the absence of a geochemical response proves that these areas have no economic potential. This is a false assumption since such a survey technique is inappropriate for these conditions.

It is the experience of the writers of this manual that, today, the majority of geochemical surveys are carried out mechanically with little thought of the suitability of the techniques employed. Frequently a stock "recipe" is used— "B" horizon soils, sieve to minus 80 mesh, analyses for so many elements, look for the high numbers. Relatively little effort is required to significantly increase the effectiveness of your surveys.

CHOICE OF METHODS

From the outset, the selection of any technique is dependent on the mineralogy and geochemistry of the target being sought. The composition of a mineral body will determine the elements that can be used. Copper is clearly ideal for a copper deposit, but for arsenic to be useful in a gold search it must be present in the gold mineralization. Mercury is only useful for mercury-bearing bodies of mineralization, etc. etc. Further, the mineralogy of the target, in combination with the secondary environment, will determine the mode of dispersion. For example, copper dispersion is both hydromorphic and mechanical while tin, typically, is almost entirely mechanical as grains of cassiterite, with a further contribution from tin in biotite and other accessory minerals.

The second point to consider is the relative disposition of the target. This may be characterized (Figure 1.1) as (1) outcropping ore, (2) partially outcropping ore, (3) buried ore concealed by younger cover, and (4) blind ore bodies completely concealed within their host rocks. Clearly, different techniques are required for these various conditions. Direct surficial sampling will be effective in cases (1) and (2) although rather different geochemical responses should be anticipated. Cases (3) and (4) demand optimized techniques that will see through cover, search beneath cover, sniff gases seeping from mineralization, detect leakage or identify halos in the surrounding rocks.

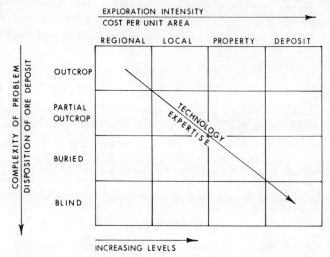

FIGURE 1.2—Some factors influencing the choice of mineral exploration survey techniques.

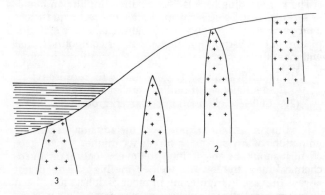

FIGURE 1.1—The variable disposition of mineral deposits with respect to the present day daylight surface. For explanation see text.

Then there is the scale of application. Geochemical surveys are used at various stages in mineral exploration with four levels of intensity readily appreciated:

(1) Regional—recognition of mineral belts or mining camps.
(2) Local—identification of targets for evaluation.
(3) Property—defining the limits of mineralized ground.
(4) Deposit—location of individual ore bodies.

As shown in Figure 1.2 these also represent increasing levels of effort, complexity of problem, sophistication of technology, expenditure per unit area and, frequently, the sequence of an exploration program. Techniques applicable at the regional scale, with broadly spaced samples, are incapable of providing the definition necessary at the property and deposit scale. Conversely, the very detailed sampling and sophistication of technology used to locate individual ore bodies are far too tedious and costly to apply at the local and regional scale.

Ultimately there is the need to integrate geochemistry with the overall exploration strategy. Depending on the character of the target, level of application, surficial environment and availability of personnel, geochemistry may take a lead or supporting role in an exploration program. Occasionally it may have no place at all. Frequently it is misapplied with surveys run to see what happens, to satisfy financial or assessment work commitments or simply because they are always run. Such surveys are very rarely truly successful.

How, then, are geochemical surveys optimized within a particular exploration program?

OPTIMIZING SURVEY TECHNIQUES

To optimize a geochemical survey a variety of techniques are available to the geochemist. The collective experience of some 50 years of exploration geochemistry can be brought to bear on the problem at hand. It is, however, necessary to be quite clear about what is wanted in an individual survey. The key feature is identified by Hawkes and Webb (1962) as RELIABILITY, which refers to the probability of obtaining and recognizing indications that an orebody is/is not present within a survey area. In constructing a reliable survey method, four characteristics are deemed desirable and represent the Basic Survey Objectives.

Basic Objectives

Optimum target identification

A target, if present, should be clearly visible in the geochemical data. It may be characterized by an increase or decrease in abundance of certain elements or a diagnostic association of elements. Regardless of the details, it should be easily distinguishable from the remainder of the survey data, a feature achieved by -

Maximum geochemical contrast

Generally the presence of mineralization is revealed by an increase in the relative abundance of the ore elements or associated guest and indicator elements in the sample material collected in the survey. Contrast is the difference between the relative abundance of an element related to mineralization and its abundance in adjacent normal, unmineralized, background situations. Contrast in soil samples is dependent on (1) the primary contrast between mineralized and unmineralized rock, (2) the relative mobility of elements in the secondary environment, and (3) dilution by barren, unmineralized material.

Sample collection, sample preparation and the choice of analytical method can all affect contrast.

Obtaining optimum anomaly contrast begins in the field through recognition of local environmental circumstances that will affect dispersion processes; sites that may be leached or enhanced due to seepage, presence of secondary precipitates, abnormal soil developments, distribution of transported overburden, etc. Field notes are thus an essential part of any survey and are reincorporated with the analytical data to aid and qualify the interpretation.

Sample collection, as stated earlier, is the most important step in the entire survey procedure. Under ideal circumstances maximum contrast is obtained by collecting a sample of any soil material directly over mineralization. These circumstances are approached in the example from Turkey provided by Koksoy and Bradshaw (Figure 1.3). In this case soils are developed in situ from the underlying bedrock (residual soils), and there has been virtually no soil horizon development with its associated chemical or mechanical reorganization. Contrast between mercury values over mineralization and background at the end of the traverse line are similar at all the depths sampled. Lateral dispersion is taking place, however, and the pattern of elevated mercury values associated with mineralization is wider at the surface than at deeper levels.

A more typical situation is found in the case history reported by Bradshaw et al. (1979), shown in Figure 1.4, from British Columbia, Canada. In this environment of high rainfall and dense vegetation there is active soil formation and strong horizon development with attendant leaching and reprecipitation of metals. Unfortunately an initial soil survey was conducted incorrectly. Samples were collected at a constant depth, generally from the base of the A horizon, which is strongly leached of all metals. This is not just a case of poor contrast, there are no meaningful geochemical patterns at all. Fortunately the problem was recognized and the area resampled with material collected uniformly from the B horizon of the soil, the top of which varies in depth from 30 to 50 cms. Results are shown in Figure 1.4. Effective geochemical mapping with good contrast that can be interpreted meaningfully has now been achieved. It is most important to sample from a constant medium (soil horizon) rather than a constant depth.

Problems can arise, however, when the character of the parent material from which the soil is derived changes across a survey area. Of particular consequence is the presence, particularly the variable presence, of exotic transported overburden. At Island Copper, British Columbia, Canada, sampling of the B horizon of soils developed on a shallow lodgement till of very local origin provides good contrast between mineralized and unmineralized locations (Table 1.1). However, where the soils are developed from transported sands and gravels (stratified drift), the copper content of the B horizon reflects that of its parent, the drift, and not the underlying mineralization. In this case contrast is lost because of the masking effects of the transported overburden.

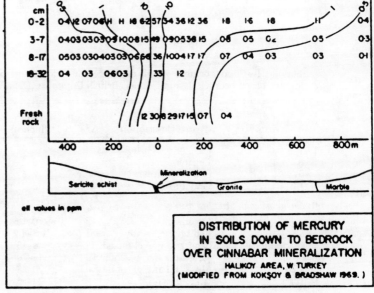

FIGURE 1.3—Distribution of mercury in residual soils down to bedrock over cinnabar mineralization. After Koksoy and Bradshaw, 1969.

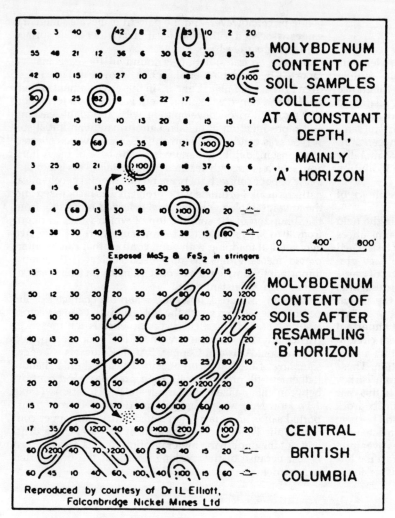

FIGURE 1.4—Molybdenum distribution in soils as shown by sampling at a constant depth (mainly A horizon) and uniformly from the B horizon. From Bradshaw et al., 1979.

TABLE 1.1—Copper content of soils developed on local till and exotic stratified drift over mineralized and unmineralized bedrock, Island Copper, British Columbia. Data from Sutherland–Brown (1975).

Horizon	Description	Depth (inches)	Cu (ppm) Unmineralized	Cu (ppm) Mineralized
Parent material: Till				
A	Organic	0–2	40	90
B	Red/gray brown	2–6	45	420
Till	Gray brown	6–8	50	1680
Parent material: Stratified drift				
A	Organic	0–1.5	40	50
B	Red brown, sandy	1.5–4.0	40	50
Glacio-fluvial	Sand and gravel mixed	4–13	50	70
Mixed	Mixed till and glaciofluvial	13–20	60	450
Mixed	Mixed till and glaciofluvial	20–28	50	170
Till	Gray brown	28–55	40	5000

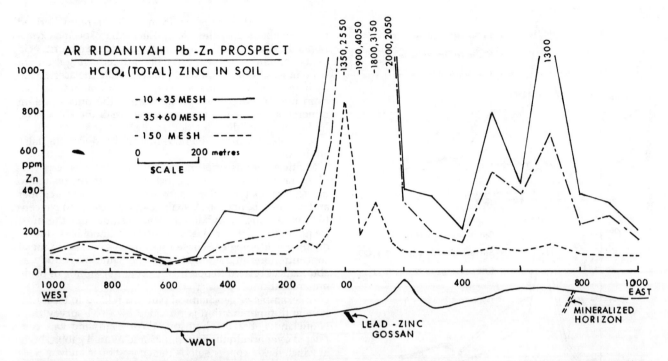

FIGURE 1.5—Horizontal distribution of perchloric acid-extractable zinc as a function of size fraction in soils over lead-zinc mineralization, Ar Ridaniyah, Kingdom of Saudi Arabia. Data from Thomson (1978) with permission from Riofinex.

Can you suggest a sampling procedure that would maintain geochemical contrast at all locations in this survey?

Attention to sample preparation can often enhance contrast. For example, Thomson (1978) demonstrated that analysis for zinc in the −10+35 mesh fraction of soil material collected at a depth of 20 cm from a semiresidual regosol in the Saudi Arabian desert (Figure 1.5) provides optimum contrast over a body of zinc mineralization. By way of comparison, data for Zn in the minus 150 mesh fraction of the same soils reveals dilution by barren aeolian material, which has severely reduced both dispersion and contrast.

Secondary enrichment of metals dispersed hydromorphically tends to occur preferentially with the fine fraction (clay and silt) of soils or as loose coatings on coarser particles. In many locations separation and analysis of the very fine fraction of soils will enhance patterns due to this process.

Geochemists have also found that contrast can sometimes be enhanced by isolating the heavy mineral fraction of soils. This is most appropriate for elements that are normally dispersed mechanically such as gold, tin and tungsten but may apply, under the right circumstances, to any element of interest that occurs in a resistate detrital form. Panning out gold grains and counting the number of grains or colors has been used by prospectors for centuries. Theobald and Allcott (1973) have used a similar approach to gain maximum contrast in the search for tin and tungsten in Saudi Arabia as have Szabo et al. (1975) in New Brunswick, Canada. More recently Farrell (1984) reports using the heavy mineral fraction of soils to improve contrast when exploring for nickel-copper-cobalt mineralization in Western Australia.

The premise behind the use of heavy mineral concentrates to optimize geochemical contrast is very simple. Can you define it?

Analytical techniques can also enhance contrast. Pluger and Friedrich (1973) showed that for soils in the environment of their investigation, geochemical contrast for fluoride changed considerably with the analytical extraction employed (Figure 1.6). Generally speaking total or near total techniques provide the best contrast in soil surveys although there are important exceptions—what are they?

Also of importance to an exploration survey is:

Minimum false alarm rate

Abnormal geochemical patterns that mimic mineralization can develop due to local processes in the secondary environment (scavenging, seepage, etc.) and also from metal-rich but unmineralized rock types. The usual procedures for reducing these false alarms employ (1) partial, specific or sequential analytical techniques for determining the mode of occurrence of metals in a sample, and (2) multielement analysis, which permits recognition of element associations characteristic of mineralization or unmineralized situations.

The distances over which groundwater may transport metals derived from weathering sulfides are highly variable and can give rise to geochemical patterns that can be hard to interpret. High metal concentrations due to secondary

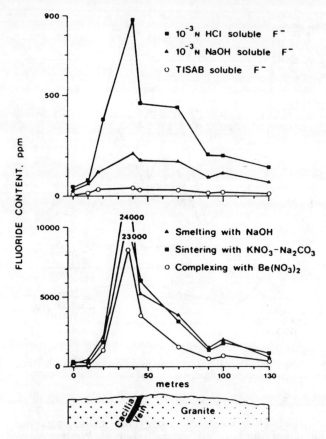

FIGURE 1.6—Fluoride distribution in soil as estimated by six different analytical extraction techniques. After Pluger and Friedrich (1973).

precipitation following hydromorphic dispersion, scavenging etc. are often characterized by weak, unstable mineral forms from which elements can be recovered by weak, partial analytical techniques. An example of this is shown in Figure 1.7 where it can be seen that the proportion of metal recovered by the weak cx extraction from soils directly over mineralization is much lower than from the nearby seepage soils.

A further feature of seepage patterns and elevated metal concentration due to hydromorphic dispersion followed by scavenging is the absence of metals that are immobile in the secondary environment. Multielement analysis can be used to differentiate residual patterns from secondary seepage patterns. For example, copper and zinc are often leached from the area of mineralization and reprecipitated in nearby seepages, whereas lead remains immobile and is anomalous only in immediate proximity to mineralization.

Finally, seepage anomalies have a characteristic topographic situation and may be interpreted from maps.

Specific or selective extraction techniques are of particular value in isolating geochemical responses related to mineralization from patterns of similar form caused by metal rich, but not mineralized, rock types. The presence of uranium in accessory minerals such as uranothorite and uranyl silicates can be distinguished from labile uranium derived from pitchblende by the use of differential, specific extractions. In Table 1.2 the hydrofluoric/nitric/perchloric acid extraction and nitric acid extraction are removing uranium from most, if not all, mineral phases, while the acetic acid/hydrogen peroxide extraction is specific for uranium derived from pitchblende. Since pitchblende is the only uranium mineral of economic importance, the use of the acetic acid/hydrogen peroxide extraction renders a survey specific to a particular form of mineralization and reduces the false alarm rate.

Multielement association and metal ratios can be particularly useful in identifying geochemical patterns related to particular rock types. Basic intrusives rich in nickel, copper and chromium, and black shales carrying elevated concentrations of zinc, molybdenum and uranium are the most frequently encountered source of false anomalies due to distinctive lithologies. These rocks often give rise to metal concentrations in soils that have the same magnitude as abundances related to potentially economic sulfide or oxide mineralization.

An example of geochemical patterns related to both rock type and mineralization is provided by a soil survey from Niquelandia, Brazil (Thomson, 1976). Sampling was conducted over an ultrabasic complex and altered gabbro body in which nickel-copper sulfide mineralization is known. Soils in this area are deep residual latosols that show virtually no vertical redistribution of copper and nickel. The results (Figure 1.8) for total copper and nickel in soils reveal a very strong and areally extensive response over the ultrabasic rocks and much smaller and generally weaker patterns related to the sulfide occurrences. On the basis of individual metal abundances and distribution patterns, it is not possible to distinguish the barren ultrabasic rocks, in which copper and nickel are most probably held in the lattice of silicate minerals, from the nickel-copper sulfide mineralization. In the Niquelandia area, the sulfides are high in copper with respect to nickel. This contrasts with the ultrabasic rocks in which nickel is high with respect to copper. Consequently, a change in the copper:nickel ratio clearly identifies the known sulfide mineralization and leads to recognition of a further target area with similar geochemical expression to the west within the ultrabasic rocks.

In addition to optimum target identification, maximum geochemical contrast and minimum false alarm rate, a survey must also display:

Cost effectiveness

Much confusion exists between the absolute cost of a survey and cost effectiveness. Project managers, ever cost conscious, are frequently tempted to pick the cheapest possible methods. This may be false economy since the techniques applied, while inexpensive, may not be effective. Frequently a small increase in cost to optimize sampling, sample preparation or analysis can dramatically improve the effectiveness of a survey. The converse is also true. Sophisticated, and expensive, techniques may promise advantages but should be critically evaluated since they may not be appropriate for the target, the local secondary environment, scale of operation or fit the overall exploration program.

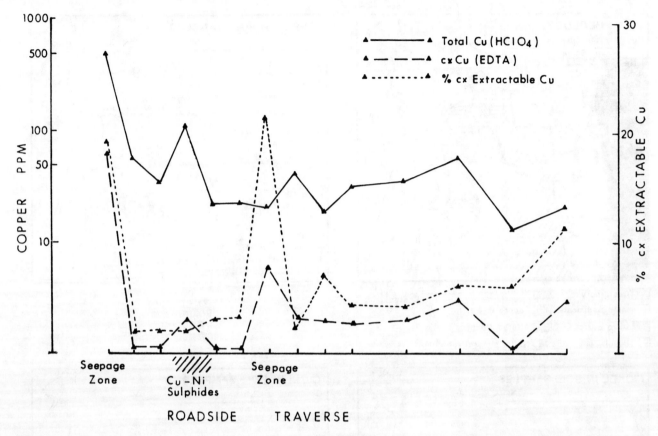

FIGURE 1.7—Horizontal distribution of perchloric acid-extractable, EDTA (cx) and percentage cx-extractable copper in minus 80 mesh, B horizon soil, Limerick Prospect, Ontario. Modified from Thomson (1975).

TABLE 1.2—The uranium content of selected rock samples as indicated by a range of analytical extraction techniques. All values in ppm. Data from Barringer Research Ltd., with permission.

Rock type	Decomposition—extraction		
	$HF + HClO_4 + HNO_3$	$4N\ HNO_3$	$H_2O_2 + HAc$
Trachyte porphyry with leucite syenite and bostonite	12.0	4.4	0.4
Leucite syenite with uranothorite	40.0	24.0	2.6
Argillite with quartz, uranyl silicates and pyrite	3100.0	1100.0	146.0
Granite with pitchblende and pyrite	400.0	240.0	440.0
Albite syenite with chlorite, hematite and pitchblende	1640.0	1080.0	1600.0

Survey Parameters

Now that we know what we want from a geochemical survey, the challenge is to design an effective program. In practical terms this means making decisions about the selection of:

(1) Sample material
(2) Sampling pattern
(3) Sample preparation
(4) Analytical procedure
(5) Criteria for interpretation of the results

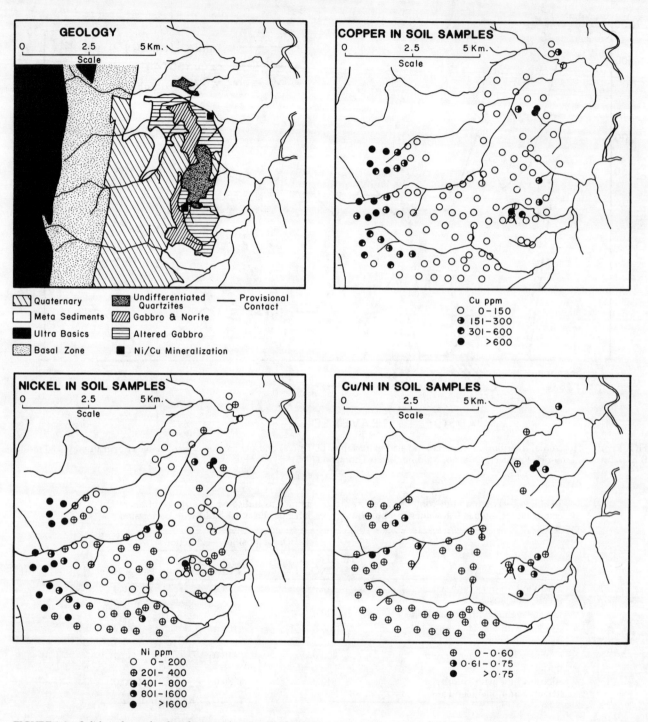

FIGURE 1.8—Solid geology, the distribution of copper, nickel, and copper:nickel ratios in near surface soil samples, Niquelandia, Goias, Brazil. After Thomson (1976). (Reproduced with permission of the Minister of Supply and Services Canada.)

To make any decision requires some knowledge or sensible assumptions about what is happening in the survey area. This means reference to relevant information on:

(1) Dispersion and mobility characteristics of elements in the mineralization and host rocks.

(2) Local environmental influences on dispersion processes.
(3) Target size—both the size of the mineralization and the expected size of any dispersion halo around it.

(4) Availability of sample material.
(5) Analytical capability.
(6) Logistical conditions.

Some general comments can be made about these various influences or survey parameters.

Element dispersion and mobility characteristics will influence the size and amplitude of secondary dispersion patterns in soils and other media. By and large, mechanical dispersion tends to produce high contrast patterns of limited dimensions modified by down slope or down ice displacement and dilution by barren material. Hydromorphic dispersion tends to give rise to broader lower amplitude patterns close to mineralization but can also produce secondary, displaced features similar to or even with higher contrast than the patterns close to mineralization in springs, seepages, bogs and nearby streams and lakes.

Local environmental influences on dispersion processes are of profound consequence. In addition to the more obvious effects due to climate and topography, the single most important factor is the parent material of soils within a survey area. Are they residual or developed on some transported material? If transported of what origin—local colluvium, alluvium, glacial till or stratified drift? Exotic materials, particularly stratified sediments, alluvium, fluvial sands and gravels, lacustrine deposits and volcanic ash, mask the bedrock. Soils developed on these materials are most unlikely to carry any simple geochemical expression of the solid geology immediately beneath.

Target size will, of course, influence selection of sample interval. Similarly, any preferred orientation of the target should be considered in the layout of sample traverses and grids. Ideally, a soil sampling grid should be set out with the baseline parallel to the long axis of the target. Cross lines are thus oriented perpendicular to the preferred orientation of the target to provide maximum opportunity to intersect it. Cross lines should be spaced such that a minimum of two adjacent lines will cross the dispersion pattern associated with a target. Similarly, sample intervals along the lines should be such that a minimum of two adjacent samples will be within the geochemical dispersion pattern of any target. Variations in the character of a geochemical response with changes in sample interval are shown in Figure 1.9 based on a uranium survey reported by Hoffman (pers. comm.). Critical examination of Figure 1.9 should permit you to see how undersampling can lead to uncertainty in interpreting a survey. Conversely, oversampling is possible for, while the geochemical pattern is very clearly and precisely revealed, is it worth the additional cost or, rather, is the cost of an entire survey at that density warranted?

What do you consider to be an optimum sample layout for this case history?

The ideal geochemical survey is based on regular, systematic sampling of the same material across a survey area. This will provide a homogeneous database within which comparative evaluation of geochemical features is possible. It is thus essential that the sampling medium selected for a survey be uniformly present across the area. Frequently this ideal cannot be obtained. Even in soil surveys the character of the soil, horizon development and differentiation will vary across a grid. In mountainous areas it is not uncommon to pass from forest to rocky scree to bog to alpine meadow over very short distances. What do you sample in each of these situations?

Analytical capability is rarely a problem in North America where we are well served by a large number of high calibre commercial laboratories. What is necessary, however, is selection of both the sample preparation and analytical method most suitable for the survey. As discussed earlier, this decision should be made in favor of RELIABILITY and provide optimum, high contrast, target identification.

Finally, logistical constraints must be evaluated. Property access, terrain conditions and availability of personnel, together with budget and time considerations, must be carefully weighed along with the other selection criteria. It may be necessary to modify sampling schemes and/or the choice of sample material in the light of logistical constraints. Should it be necessary to reduce the field commitment, the survey may be modified by taking advantage of enhanced dispersion of an indicator element to reduce sample interval and hence the number of samples. Alternatively, in high cost field areas it may be prudent to capitalize on the time in the field, and ultimately reduce large logistical costs, by oversampling and analyzing only every second or third sample. Definition of interesting geochemical features can be achieved by analyzing the intervening samples. In effect a second fill-in survey has been possible without the cost of mobilizing to the field.

Full optimization of a survey is, however, only possible with access to information relevant to the survey area. This

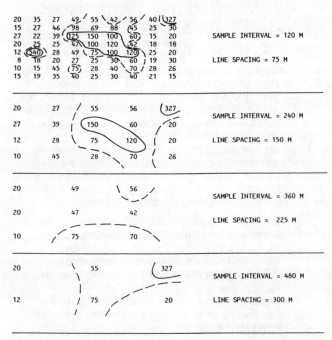

FIGURE 1.9—Variation in the character of a geochemical response with change in sample interval on a soil grid. From Hoffman, personal communication.

is obtained by conducting Orientation Studies that will provide the technical information on which to base operational procedures.

Orientation Studies

Orientation studies are best described as a series of preliminary experiments to determine the character of geochemical dispersion related to mineralization in a given location. The information obtained is used to: (1) define background and abnormal geochemical responses, (2) define optimum survey procedures, (3) identify those factors that influence dispersion and are thus criteria for the interpretation of survey results, and (4) recognize those features that must be noted and reported by the samplers. An orientation can be conducted, with varying effectiveness, in one of three ways.

The orientation survey

A classic orientation survey comprises field sampling and analysis around a representative body of known mineralization. Procedures for such surveys are fully described by Hawkes and Webb (1962) and Bradshaw (1975). Ideally, the work is conducted over mineralization at a location that is geologically and geomorphologically representative of the proposed survey area and is continued well away from mineralization to adequately define background conditions. Bradshaw (1975) recommends the following procedures for soil surveys.

Orientation soil samples should be collected from at least two traverses across mineralization and continue well into background. Sample spacing depends on the extent of the mineralization, but at least four or five samples should be collected over it and also background. It is important that the character of different soils be evaluated. As a result the traverses should cover all normal physiographic conditions and major soil types encountered, such as well-drained ground, steep slopes, seepage areas and bogs. Variation within and between soil horizons and with depth should be tested. Samples should be collected from every recognizable soil horizon, or at 20 cm intervals down the profile, whichever is less.

In addition to the near universal minus 80 mesh (177 microns) fraction, various other size fractions of the sample material should be prepared and representative analytical procedures applied to all the subsamples. We recommend that soils be dry sieved into the following fractions:

Mesh (ASTM)	Aperture (Microns)
$-35 + 80$	$-500 + 177$
-80	-177
$-80 + 140$	$-177 + 105$
$-140 + 230$	$-105 + 63$
-230	-63

Bradshaw (1975) also recommends preparation of a heavy mineral fraction if detrital resistate dispersion is suspected and certainly if gold, tin, tungsten, tantalum or similar metals or minerals are involved.

All samples should be analyzed using a total extraction technique. In addition it is recommended that soil samples be analyzed by a hot acid extractable and cold extractable technique as well as any specialized (such as sulfide specific, tin specific, organic matter specific) technique that may be desirable.

A literature study

It may be quite impractical to visit a field area and conduct an orientation survey ahead of the main exploration program. Under such circumstances much significant information of consequence to the design of a survey may be obtained by review of other people's work. These may be published papers or assessment reports and other internal company documents. Frequently it is possible to conduct a reverse orientation by critically evaluating historical survey results and noting both the successes and limitations of the work. Literature surveys are usefully complemented by discussion with people knowledgeable of conditions in the survey area and consultation with professional geochemists.

A theoretical orientation

This highly speculative approach is based on application of theoretical models, the basic principles of geochemistry and assumptions as to the solid and surficial geology, geomorphology and climate of the survey area. Although sometimes successful, this procedure is to be discouraged and is almost as despised as the unranked, unnumbered, unmentionable fourth procedure—No Orientation!

By now you should realize that you can do much better than the sheep in industry.

SURVEY ORGANIZATION AND OPERATION

We do not propose to discuss the actual management of a geochemical survey in exhaustive detail. A check list of points for consideration is provided in Table 1.3, which draws attention to some of the problems we have seen develop from time to time. Essentially we recommend careful and patient attention to detail. Murphy's Law—"anything that can go wrong will go wrong"—is particularly true for geochemical surveys as will emerge from the case histories presented in this book.

However, assuming you have been wise and have chosen a practical orientation study to define survey parameters, there are a few last things we believe you must do before the routine work begins. When the field crew is mobilized to the field area, you, too, should be there to:

(1) Show the samplers exactly what you want collected and to train them in survey procedures.
(2) Examine for and confirm the character and distribution of any transported overburden.
(3) Verify soil conditions at key locations in the survey area.
(4) Familiarize yourself with the physiography of the survey area in preparation for the later interpretation phase.

So much for the rational and thought processes behind the design of a survey. Let us now tackle some real examples.

TABLE 1.3—Check list for the organization of a geochemical soil survey.

Item	Check
FIELD PARTY	numbers, composition, experience, leader
TRAINING	when, where, by whom
BASE MAPS	appropriate scale, topography, etc.
NUMBERING SCHEMES	simple unambiguous, avoid complex alphanumerics
FIELD NOTES	make sure they are taken correctly
QUALITY CONTROL	collect field duplicate samples and insert, with standards, in batch submitted to laboratory
COMMUNICATIONS WITH LABORATORY	must be simple and direct. Only designated personnel should actually give instructions to the laboratory.
SHIPPING LISTS	must accompany every consignment sent to the laboratory
INSTRUCTIONS	give clear unambiguous instructions to the laboratory
RETURN OF DATA	check duplicates, standards, etc. for quality of analytical data, request reanalysis when in doubt
DATA HANDLING	manual or computer aided. What procedures are best for your project?
INTERPRETATION MAPS	prepared to summarize geochemical features
INTEGRATION OF FIELD NOTES	used to qualify interpretation of geochemical data
STORAGE OF DATA	need to be able to retrieve for reinterpretation
ARCHIVE OF SAMPLES	at laboratory, office or warehouse
INTEGRATION WITH OTHER EXPLORATION PROCEDURES	ensure good communication with management and other project personnel
REPORTING	author of report *must* be familiar with field program

PROBLEM 1: MOLYBDENUM IN SIERRA LEONE

Objective

You have been assigned to a mineral exploration program in West Africa where a company has already established its land positions. A number of mineral occurrences have been located on the properties but little systematic evaluation of the ground has been undertaken.

The planned program is to first screen the ground around all known occurrences by a combination of techniques including geochemistry and then complete a more detailed appraisal (up to drilling) of properties with the highest potential. You are responsible for establishing optimum geochemical techniques for the project.

Description of the Area

The survey area is underlain by deformed metasediments and metavolcanics enveloped in granitic gneiss. The rocks are schistose and of variable composition including amphibolite, chlorite and talc schists, quartzites and ironstones intruded by younger granites.

One of the known mineral occurrences is the site for your orientation. At this locality a number of polymetallic veins carrying pyrite, chalcopyrite, galena, molybdenite and sphalerite are found in amphibolite and granite. The occurrence is believed to be subeconomic in size. The veins form pods in fractures, and none have been traced for more than 90 feet along strike.

The survey area is a lightly dissected plateau characterized by gentle slopes and a deep residual overburden up to 60 ft. thick; soils are typically lateritic. The area is within the humid tropics and is covered by tropical rain forest, which gives way to open grassland and forest along ridges.

Access to the area is by road, and within the area there are numerous footpaths.

Assumed

Certain decisions have been made before embarking on the orientation survey.

(1) Because of the size of individual properties (maximum 10 square miles), soil sampling was selected as the most cost effective geochemical method for screening the prospects.
(2) You have learned that other people working in similar areas have been using conventional soil sampling techniques—apparently with success. They have sampled the B soil horizon and analyzed the minus 80 mesh fraction obtained by dry sieving. The orientation attempts to challenge this approach and provide data that will allow you to optimize your survey.
(3) For the purpose of this exercise, data for only one element, molybdenum, will be considered.
(4) You are obliged to use the facilities of a local laboratory, which can only perform molybdenum analyses by one technique; a colorimetric determination using zinc dithiol following an alkali fusion digestion.

Questions

Results of the orientation, which comprised sampling at several points on the showing, are given in Tables 1.4–1.7. Examine the data critically and then proceed to answer the following.

TABLE 1.4—Molybdenum content of minus 80 mesh fraction of residual soils over mineralized and barren rock. Data from Mather (1959).

Bedrock type	Number of samples	Mo ppm Range	Mean
Mineralization at A	4	1200–2200	1500
Mineralization at B	3	90–120	100
Synkinematic granite	10	1–5	1.9
Late-kinematic granite	12	1–10	4.0
Amphibolite	36	1–3	2.0
Talc schist	4	1–2.5	1.5
Pelitic sediments	29	1–3	1.6
Bedded ironstones	7	1–5	2.6

TABLE 1.5—Variation of molybdenum content of minus 80 mesh fraction of soils with depth. Data from Mather (1959).

Location	Horizon Surface	A1	A2	B
Over mineralization	30	60	100	120
200 ft from mineralization	10	15	35	45
400 ft from mineralization	10	10	15	20
Background	4	3	4	2

1. What are the general mobility characteristics of molybdenum in the secondary environment (weathering mineralization and host rocks, humid tropics, deep weathering, lateritic soil) of the survey area?
2. What local situations will further affect the mobility and/or concentration of molybdenum in the secondary environment?
3. What is the character (total, partial, etc.) of the analytical technique for molybdenum that you are obliged to use, its limitations and advantages; and what constraints, if any, will these features place on the interpretation of any subsequent survey data?
4. In the light of (1), (2) and (3) above, what field observations should be made and/or additional analytical parameters (elements, etc.) determined?
5. What is the optimum soil horizon for sampling?
6. Is the minus 80 mesh fraction a suitable medium for routine surveys?
7. Can you recommend a better choice of size fraction as the medium for (1) an initial evaluation of the properties, and (2) a detailed study of individual prospects?
8. What are your recommendations for the layout (interval) of soil samples during the various phases of the exploration program?
9. What procedures would you introduce to verify the quality and efficiency of the survey program?

For further background on this case history and answers to some of the questions, you are referred to the work of Tooms et al. (1965).

PROBLEM 2: EXPLORATION FOR BASE METALS IN A GLACIATED AREA OF CENTRAL NORWAY

(Contributed by M. B. Mehrtens)

Objective

The Hjerkinn–Folldal district of central Norway is the site of a number of known bodies of massive sulfide mineralization and has the geological potential for more. In this case history we pick up the story several years ago when it was considered necessary to look critically at the use of geochemical methods in this area.

Early geochemical studies by the Norsk Geologiske Undersokelese (NGU) demonstrated, among other things, that there was no appreciable mechanical (ice) dispersion of metallic sulfides despite the fact that one of the earliest discoveries had been made as a result of finding boulders of ore. Accordingly further work was undertaken (Mehrtens, 1966; Mehrtens and Tooms, 1973) to determine the major mechanisms by which base metals are secondarily dispersed from the massive sulfide deposits beneath glacial overburden and thus provide for improved techniques of geochemical exploration in this region.

Description of the Area

The district is characterized by rolling hills, broad U-shaped valleys and elevations of between 700–1,400 m above sea level. The climate is cold and dry: mean annual temperature at Hjerkinn is $-0.8°C$ and mean annual precipitation 230 mm. Vegetation above the tree line is dominated by dwarf species of willow, birch and juniper with predominantly coniferous forest below the tree line. The region was glaciated by the Pleistocene continental ice sheet and is now mantled by glacial drift comprising lodgment and ablation till with subordinate glacio-fluvial material. The glacial overburden is of highly variable thickness, but commonly in the range 3–4 m. Bedrock and lodgment till within the district are siliceous, and the soils are poorly developed podzols. Mineralization within the study areas consist of lensoid bodies of massive sulfide associated with the Lower Ordovician

TABLE 1.6—Molybdenum content of different size fractions of B horizon soils. Data from Mather (1959).

Location	Size fraction (ASTM mesh)						
	$-10+20$	$-20+38$	$-38+80$	$-80+125$	$-125+197$	-197	-80
Over mineralization	200	150	80	100	90	140	120
200 ft from mineralization	10	20	15	35	40	50	45
400 ft from mineralization	4	3	6	8	6	20	20
Background	4	3	2	3	2	2	2

TABLE 1.7—Distribution of molybdenum in minus 80 mesh fraction of soil from two traverses across mineralization. Data from Mather (1959).

Station	Mo ppm	Remarks	Mo ppm	Remarks
+2500 E	3	Bedrock amphibolite	3	Bedrock amphibolite
+2000	4		4	
+1900	3	♦ River	3	
+1800	15		4	
+1500	50		5	
+1000	40		6	
+ 800	60		5	
+ 600	90		7	
+ 500	60		6	
+ 400	130	Ground slopes	20	
+ 300	150	East at 10'	35	Slope 5'
+ 200	170		45	East
+ 150	450		40	
+ 100	225		80	
+ 75	800		70	
+ 50	1700		90	
+ 25	1200		100	
+ 00	2200	Vein	120	Mineralization
+ 25 W	1300		85	Bedrock
+ 50	500		40	Granite
+ 75	100		15	
+ 100	90		20	Flatground
+ 150	55		10	
+ 200	70	Ground slopes	10	
+ 300	30	West at 5'	8	
+ 400	45		5	
+ 500	20		5	Slope 5'
+ 600	40		3	West
+ 800	15		2	
+1000	5		2	
+1500	3		2	
+2000	4		2	

Storen volcanic group within the Caledonian eugeosyncline.

A number of mineralized localities and background areas were examined in the Hjerkinn–Folldal district. Of these Tverrefjellet is at a freely drained location above the tree line, while Sondre Gjeitteryggen is at a freely drained site below the tree line.

The background metal values in the soil horizons and underlying till are summarized in Tables 1.8 and 1.9. Greater detail on the metal distribution patterns related to mineralization are given in Tables 1.10 and 1.11 and Figures 1.10–1.22.

As indicated in Figure 1.11 secondary iron oxides at Tverrefjellet cement the till overlying and downslope of the ore deposit. There is also a collapsed area and swallow hole in the suboutcrop of the massive sulfide body. Down drainage of this swallow hole the drift is heavily cemented with secondary iron oxides, and groundwater emerging in seepages within the iron oxide zone was found to contain large concentrations of dissolved metals at a pH of 2.5 (Table 1.10).

Assumed

The following information on the methods employed in the study should help understanding of the results.

(1) All soils data are from analysis of the minus 80 mesh fraction obtained by dry sieving.
(2) Copper data are obtained as follows. Results designated Cu are from a $KHSO_4$ fusion extraction technique that provides what is effectively

TABLE 1.8—Geochemical values in the glacial drift and overlying soil horizons in profiles from background areas, Hjerkinn (data on the minus 80 mesh fraction). Mean values from 10 profiles. Data from Mehrtens and Tooms (1973).

Depth (cm)	Description	ppm			cxCu (ppm)
		Cu	Zn	Pb	
0–5	black-brown organic	5	70	10	0.2
5–15	gray leached loam	10	60	10	0.5
15–25	brown medium-grained loam	20	100	10	0.7
25–30	semidecomposed	20	100	10	0.6
30–50	gray	30	100	10	0.6
50–90	gray	50	150	10	1.0
90–120	gray	60	160	10	1.1
120–150	gray	60	140	10	1.2
150–180	gray	65	160	10	1.2
170–180	gray	65	160	10	1.2
180–200	gray	70	160	10	1.2

TABLE 1.9—Calculated background and threshold values for the solum and moraine in the Hjerkinn-Folldal area. Data on minus 80 mesh fraction: all values in ppm. From Mehrtens (1966).

Element		Soil (n = 50)	Till (n = 70)
Cu	Mean	15	40
	Range	2–35	10–90
	Threshold	30	80
Zn	Mean	70	80
	Range	30–155	20–230
	Threshold	140	180
cxCu	Mean	0.5	0.6
	Range	0.2–1.1	0.2–1.6
	Threshold	1.0	1.4

TABLE 1.10—Properties of filtered groundwater, Hjerkinn area, central Norway. Data from Mehrtens (1966).

Locality	pH	Metal content (ppb)			
		Cu	Zn	Pb	Fe
(a)	2.5	18,400	23,000	640	360,000
(b)	3.8	69	214	6.8	18,000
(c)	6.8–7.3	3–9	18–30	0.3	188–260

(a) Immediately downslope from the Tverrefjellet ore deposit.
(b) Within Trench 1 Tverrefjellet at suboutcrop of orebody.
(c) Near-surface groundwaters in barren area, Hjerkinn.

TABLE 1.11—Cu:Zn:Pb ratios in unoxidized ore, gossan and anomalous basal till, Tverrefjellet. Data from Mehrtens and Tooms (1973).

Material	Ratio of		
	Cu :	Zn :	Pb
Unoxidized ore	4 :	6 :	1
Gossan	2.3 :	1 :	1
Anomalous basal till	40 :	25 :	1

total copper. Results designated cxCu are obtained following a weak acid decomposition that recovers copper in exchange positions, sorbed onto clays, secondary iron oxides and organic matter.

Questions

Carefully read the background information given above and then critically examine the accompanying figures and tables before answering any of the questions.

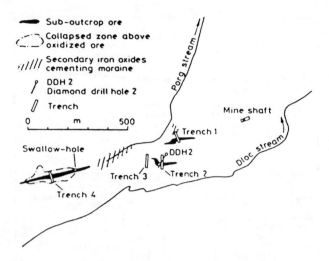

FIGURE 1.10—Surface features at Tverrefjellet. From Mehrtens and Tooms (1973).

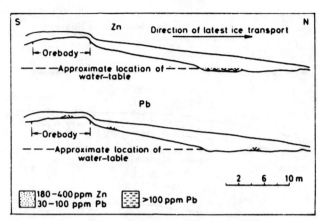

FIGURE 1.13—Distribution of Zn and Pb in overburden, Trench 1, Tverrefjellet. From Mehrtens and Tooms (1973).

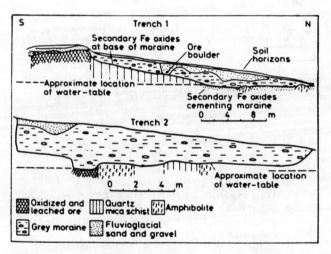

FIGURE 1.11—Geology of Trench 1 and 2, Tverrefjellet. From Mehrtens and Tooms (1973).

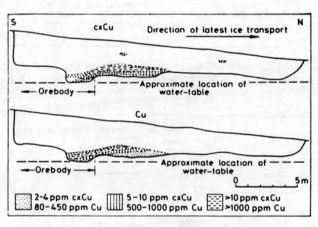

FIGURE 1.14—Distribution of anomalous cx Cu and Cu in overburden, Trench 2, Tverrefjellet. From Mehrtens and Tooms (1973).

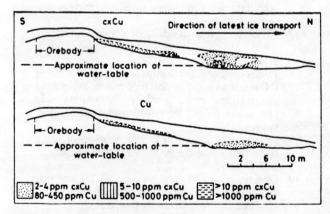

FIGURE 1.12—Distribution of anomalous cx Cu and Cu in overburden, Trench 1, Tverrefjellet. From Mehrtens and Tooms (1973).

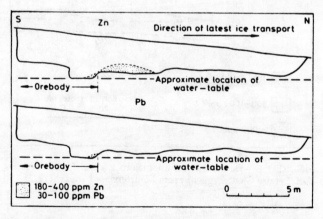

FIGURE 1.15—Distribution of anomalous Zn and Pb in overburden, Trench 2, Tverrefjellet. From Mehrtens and Tooms (1973).

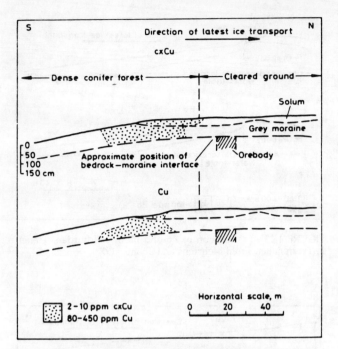

FIGURE 1.16—Distribution of anomalous cx Cu and Cu in overburden, Sendre Gjeitteryggen. From Mehrtens and Tooms (1973).

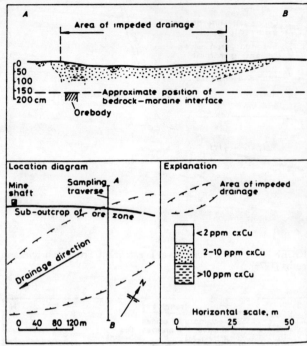

FIGURE 1.18—Distribution of anomalous cx Cu in overburden, Nordre Gjeitteryggen. From Mehrtens and Tooms (1973).

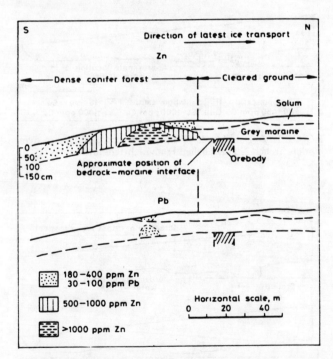

FIGURE 1.17—Distribution of anomalous Zn and Pb in overburden, Sendre Gjeitteryggen. From Mehrtens and Tooms (1973).

1. What are the general mobility characteristics of copper, zinc and lead?
2. What are their relative mobilities under conditions prevailing in the study area?
3. Describe the dominant mode of base metal dispersion at Tverrefjellet and Gjeitteryggen and your evidence for any particular process.
4. What has been the influence of the Pleistocene glaciation on base metal dispersion patterns at the two locations?
5. Is there any conclusive evidence that surface vegetation is affecting secondary dispersion processes?
6. What are your recommendations as to the preferred medium for sampling in a routine exploration survey?
7. Given complete freedom of choice, what elements and analytical techniques would you select for your survey?
8. The secondary environment is clearly a major influence on geochemical dispersion. What field observations or additional measurements would you make to aid later interpretation of your survey?
9. Do the data give you any indication of ways to further optimize geochemical techniques in this region?
10. Describe the character of any additional orientation work you believe to be desirable.
11. Can you now explain why the NGU found "no appreciable mechanical (ice) dispersion of metallic sulfides"?

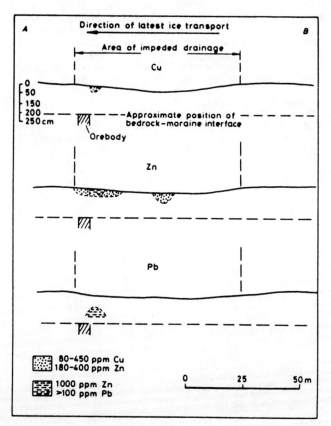

FIGURE 1.19—Distribution of anomalous Cu, Zn and Pb in overburden, Nordre Gjeitteryggen. From Mehrtens and Tooms (1973).

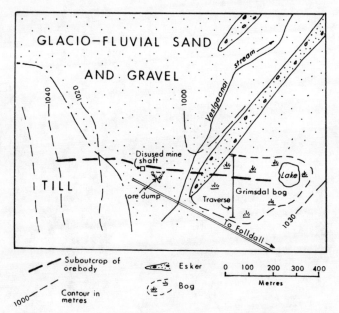

FIGURE 1.20—Geology and physical features in the vicinity of the Grimsdal ore deposit, central Norway. After Mehrtens (1966).

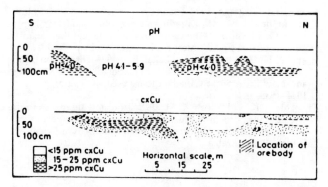

FIGURE 1.21—pH and cx Cu peat values, Grimsdal bog. From Mehrtens and Tooms (1973).

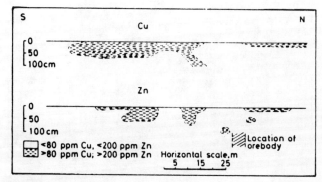

FIGURE 1.22—Distribution of anomalous Cu and Zn in peat, Grimsdal bog. From Mehrtens and Tooms (1973).

REFERENCES

Bradshaw, P.M.D. (editor) 1975. Conceptual models in Exploration Geochemistry—the Canadian Cordillera and Canadian Shield. Journal Geochemical Exploration, v. 4, p. 1–213.

Bradshaw, P.M.D., Clews, D.R. and Walker, J.L. 1979. Exploration Geochemistry. Barringer Research Ltd., Toronto, 54 pp.

Farrell, B.L. 1984. The use of "loam" concentrates in geochemical exploration in deeply weathered arid terrains. Journal Geochemical Exploration, v. 22, p. 101–118.

Hawkes, H.E. and Webb, J.S. 1962. Geochemistry in Mineral Exploration. Harper and Row, London. 415 pp.

Koksoy, M. and Bradshaw, P.M.D. 1969. Secondary dispersion of mercury from cinnabar and stibnite deposits, West Turkey. Colorado School of Mines Quarterly, v. 64, p. 333–356.

Mather, A.L. 1959. Geochemical prospecting studies in Sierra Leone. Unpublished Ph.D thesis. University of London, England.

Mehrtens, M.B. 1966. Geochemical dispersion from base metal mineralization, central Norway. Unpublished Ph.D thesis. University of London, England.

Mehrtens, M.B. and Tooms, J.S. 1973. Geochemical dispersion from sulphide mineralization in glaciated terrain, central Norway. In: Jones, M.B. (editor), Prospecting in Areas of Glacial Terrain. Institute Mining and Metallurgy, London, p. 1–10.

Pluger, W.L. and Friedrich, G.H. 1973. Determination of total and cold-extractable fluoride in soils and stream sediments with an ion-sensitive fluoride electrode. In: Jones, M.B. (editor), Geochemical Exploration 1972. Institute Mining and Metallurgy, London, p. 421–427.

Sutherland–Brown, A. 1975. Island Copper deposit, British Columbia. In: Bradshaw, P.M.D. (editor), Conceptual Models in Exploration Geochemistry—The Canadian Cordillera and the Canadian Shield. Journal Geochemical Exploration., v. 4., p. 76–78.

Szabo, N.L., Govett, G.J.S. and Lajtai, E.Z. 1975. Dispersion trends of elements and indicator pebbles in glacial till around Mt. Pleasant, New Brunswick. Canadian Journal Earth Science, v. 12, p. 1534–1536.

Theobald, P.K. and Allcott, G.M. 1973. Tungsten anomalies in the Uyaijah ring structure, Kushaymiyah igneous complex, Kingdom of Saudi Arabia. Ministry of Petroleum and Natural Resources, Jeddah, Saudi Arabia.

Thomson, I. 1975. Limerick Ni–Cu prospect, Ontario. In: Bradshaw, P.M.D. (editor), Conceptual Models in Exploration Geochemistry—The Canadian Cordillera and the Canadian Shield. Journal Geochemical Exploration, v. 4, p. 168–172.

Thomson, I. 1976. Geochemical studies in central-west Brazil (bilingual edition). D.N.P.M., Ministerio das Minas e Energia, Brazil, 258 p.

Thomson, I. 1978. Geochemical orientation studies, Kingdom of Saudi Arabia. Unpublished report. Riofinex.

Tooms, J.S., Elliott, I.L. and Mather, A.L. 1965. Secondary dispersion of molybdenum from mineralization, Sierre Leone. Economic Geology, v. 60, p. 1478–1496.

Chapter 2

GEOTHERMAL EXPLORATION—THE SOIL SURVEY

S. J. Hoffman

INTRODUCTION

Once the decision has been made to use geochemistry on a project, the resulting survey can either significantly enhance or detract from the exploration effort depending on the level of expertise applied to sample collection, analysis and interpretation of results. In this section, the search for an unconformity-related U deposit will serve as a model to illustrate general principles: the reader is then asked to apply the same concepts to exploration for an epithermal Au deposit.

PHASE 1—THE OFFICE

The unconformity-related uranium deposit

The exploration program starts in the office with a definition of exploration objectives. Most important of these objectives is a statement regarding the type of deposit to be the focus of the exploration effort—the model. Few exploration programs plan to look for U period. Rather they are organized to look for unconformity-related U, roll front U, or vein-type U, etc. If by fortuitous accident, U is found in commercial quantities in some other type of deposit, so much the better, but it would not have been the focus of the initial planning effort. It could, however, modify subsequent work by providing a new (and better?) model on which to base exploration decisions.

Unconformity-related uranium deposits are the most sought after of all U deposits. The first question to be posed by the inexperienced would be: "What is an unconformity-related U deposit?". Before answering this, it is worth remembering that differences in opinion would result if a geologist was asked the question as opposed to a geophysicist or geochemist. From an exploration viewpoint the best answer represents a consensus among the three disciplines. Experience has shown that without a general consensus, the exploration effort may be wasted through misapplication of the various technologies.

The unconformity related U model is summarized in Figure 2.1, which illustrates its main geological properties. Most ore is located in a thin zone along the unconformity between a Proterozoic sandstone and older basement rocks. For the deposit to have been preserved from erosion, a cap of sandstone must be present. The thickness of this cap affects the economic desirability of a deposit, with thicknesses in excess of 150 to 200 m increasing development and mining costs dramatically.

A geochemical model (Figure 2.2) can be suggested which assumes that, during formation of the ore deposit, uranium was also deposited along faults cutting the sandstone and in permeable horizons within the sandstone. Geochemistry becomes particularly effective when this primary halo, rather than the deposit itself, forms the target of the exploration program.

It should be anticipated that pathfinder elements, such as Cu, Ni, As and Ag and radiogenic daughter products of uranium, such as Ra, might also be present in anomalous quantities. Knowledge of the geochemistry of U, its daughter elements and potential pathfinder elements is therefore needed, particularly in the context of the environment to be explored. Relevant information may be found in Levinson (1980) or Rose, Hawkes and Webb (1980). These authors summarize average backgrounds in soil, rock, water and vegetation and give geochemical associations and elemental mobility in a variety of surficial environments as well as information on the occurrence of the element's ores and

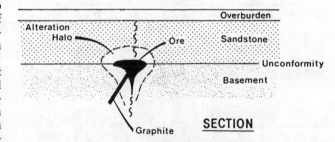

Empirical Parameters Common to many Deposits

- At or near Sandstone-Basement Unconformity
- Metasedimentary Basement Rocks
- Spacial Relationship to Graphitic Zones
- Structures with both Pre & Post Sandstone Movement
- Alteration Halos

FIGURE 2.1—Generalized Athabasca model, unconformity type of U deposit. After Hoffman (1983).

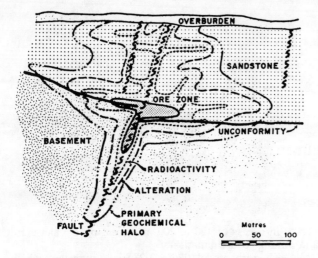

FIGURE 2.2—Hypothetical relationship between primary geochemical halo, gamma radioactivity and bedrock alteration, unconformity U model. After Hoffman (1983).

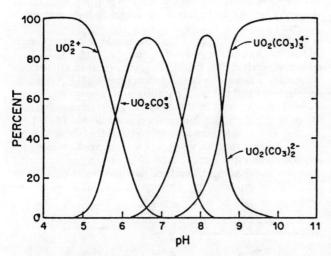

FIGURE 2.3—Distribution of aqueous uranyl-carbonate species as a function of pH for $PO_4 = 0.1$ ppm; $P_{CO_2} = 10^{-2}$ atm; $U = 10^{-8}$ M (2.4 ppb). After Levinson (1980).

minerals. Table 2.1 illustrates the contents of such a table for U. Figure 2.3 indicates that solubility of U, particularly under alkaline conditions, is likely to be a significant factor in dispersion processes and will affect the design and interpretation of geochemical surveys.

Knowledge of factors beyond those found in published tables might be necessary for understanding the behavior of some elements or if some newly developed technique is being actively employed. The recent history of the use of radon in uranium exploration seems to illustrate this point. Some users of the technique in the mid 1970's committed large expenditures to preliminary surveys in new environments hoping that anomalies would be so outstanding that the need to consider complicating factors could be avoided.

Many of these surveys ran into problems because of a lack of information on dispersion processes and disillusioned users often abandoned use of radon geochemistry. However, faced with contradictory case histories, the prudent companies continued research with the objective of eventually adding this procedure to the exploration tool box and achieving a competitive advantage. Table 2.2, for example, outlines probable controls on radon dispersion leading to formation of false anomalies in permafrost terrains.

Determination of U by the laboratory is not as straightforward as for many other elements. A number of extraction techniques are available, including a variety of partial and total extraction procedures. In Canada the most commonly used extraction has been a 4N nitric acid digestion with the final determination by ultraviolet fluorescence. An analytical problem, quenching of U fluorescence by high concentrations of Mn and/or Fe and/or base metals (Table 2.3), was not fully appreciated until about 1978. The astute explorer might therefore scan geochemical results obtained for a survey area prior to that time to determine if opportunities might have gone unrecognized. Within the United States total digestion procedures were often employed, but U was then extracted into an organic solvent prior to the fluorescence determination to avoid the quenching problem mentioned above.

Was the total digestion better than a partial 4N nitric acid extraction?

To answer this question, remember that the exploration program is searching for U in pitchblende. A total determination, however, will include U in zircon, U in apatite, U in rare earth minerals and so on. By increasing background U concentrations, U from these additional sources will lower the contrast of U anomalies related solely to pitchblende occurrences. The partial extraction would therefore be preferred in most cases. Geochemical exploration strategy must attempt to maximize anomaly signals due to the presence of mineralization relative to background variations related to sample composition and analytical factors. Determination of absolute metal abundance is not necessarily desirable.

Lastly, and again peculiar to U exploration, the relationship between radiometric prospecting and U geochemistry has to be addressed. Most unconformity U deposits have been found by discovery of anomalous radioactivity (boulders) on mapping or prospecting traverses. Orientation studies conducted around many ore deposits have indicated positive litho and soil geochemical signatures. Complementary radiometric surveys and geochemical sampling might reduce the probability of missing a potentially significant anomaly.

So far consideration has been given to geochemistry, but in the office all available exploration techniques would be given equal time in a general discussion of their suitability and cost effectiveness. Empirical relationships need to be summarized to define possible exploration options and their costs. Finally, after all pertinent facts are stated, a general discussion is needed to determine what constitutes a drill target. This will determine exploration strategy, for if anomalies are generated that can never be drill targets, at least for participants in the general discussion, they best not be

TABLE 2.1—Geochemistry of uranium (based on Levinson, 1980).

Igneous rocks (ppm): average 2.7; mafic 0.6; intermediate 3; felsic 4.8.

Sedimentary rocks (ppm): shale 4; black shale 3–1250; sandstone 0.45–3.2; limestone 2.

Soils (ppm): average 1.

Surface waters (ppb): average 0.4.

Vegetation ash (ppm): 0.5; poor biological response but associated elements such as Ra, Mo and Se may be useful.

Geochemical associations: lithophile; U, Th, rare-earths, P, F, Zr, Ti, Mo, Bi, Cu, Ag, Zn, etc., depending on the type of igneous association (e.g., pegmatite, carbonatite); U, Cu, V, Se, Mo, C in sandstone type; U-Au in placers; see below under Pathfinders for additional associations.

Ore minerals: uraninite, brannerite, carnotite.

Substitutions in: zircon, apatite (including phosphorites), allanite, niobate-tantalates (e.g., euxenite), monazite.

Soils—occurrence in: resistate, clastic and transported minerals and rock fragments; adsorbed on organic matter, clays, and iron oxides.

Secondary minerals: phosphates, vanadates, carbonates; also uraninite.

Aqueous phase: uranyl carbonate and phosphate complexes.

Mobility—primary environment: highly mobile; concentrates in late phases (granites, hydrothermal veins).

Mobility—secondary environment: highly mobile in the oxidizing environment, especially alkaline.

Geochemical barriers: reduction (Eh); adsorption; special ion precipitates (e.g., vanadates such as carnotite).

U is a pathfinder for: uranium deposits; it has potential for Au-U placers, certain Ag-Au veins, and carbonatites.

Pathfinders for U are: depending upon type of deposit: (a) sandstone or roll-front—Mo, Se, V, Cu, C; (b) classical vein (e.g., Beaverlodge, Sask.)—Cu, Ag, Co, V, Ni, As, Au, Mo, Bi, Se; (c) unconformity vein (e.g., Key Lake, Sask.)—Cu, Ag, Co, Ni, As, V, Se, Mo, Au; (d) pegmatite—Th, Mo, Nb, Ti, rare-earths; (e) carbonatite—Nb, Th, Cu, F, P, Ti, Zr, rare-earths; (f) placer (e.g., Elliot Lake, Ont.)—Th, Ti, Au, Zr, rare-earths. In addition, Rn, He and Ra for all type of U deposits.

Comments: All modern exploration geochemists must understand the factors governing the mobility of U and its daughter products (e.g., Rn, Ra) in the secondary environment, complexing and disequilibrium.

TABLE 2.2—Factors that may influence radon distribution in the surficial environment.

Factor		Comments
Radium anomalies		Geochemical enrichments of radium can be expected to cause radon anomalies. The geochemical behavior of radium is probably the most critical factor of all.
Mineralized float		Mineralized bounders or microboulders may cause anomalies unrelated to a bedrock source.
Groundwater	Seepage areas	Upward movement of groundwater in base of slope regions may result in enhanced radon readings.
	Flow regimes	Radon anomalies may be displaced downslope from their source.
	Surface slope	Steep slopes should result in greater displacement of radon anomalies than gentle slopes.
Soil & overburden	Water saturation	Degree of saturation will influence radon movement; lateral variations can cause anomalies.
	Texture	Influences radon movement in either soil gas or groundwater.
Permafrost	Presence	Can be expected to slow radon migration, depress backgrounds.
	Unfrozen layer thickness	Lateral variations may influence radon movement, e.g., frost boils. Radon migration near lakes is probably greater than away from lakes because of the moderating influence an unfrozen body of water has on depth to permafrost.

actively sought if this can be predicted in advance. Table 2.4 summarizes results of such a discussion. Note that only four drill target situations have been proposed. The project manager can proceed with the exploration program clearly knowing what must be found to achieve an effective (and harmonious) testing of anomalies.

The epithermal gold deposit

The unconformity-related U deposit was the principle exploration target in Canada during the period 1975 to 1980 and resulted in the discovery of many valuable deposits, most notably the $11 billion (1985 gross value) Cigar Lake

TABLE 2.3—Quenching of U fluorescence (from Bradshaw et al., 1979).

	Sample number	U content (ppm)	
		Direct determination	After extraction
Soils	111206	58	282
	111207	1	3
	111209	2	5.5
	111270	5	16
Lake sediments	130062	0.2	1.0
	130075	0.5	1.1
	130092	0.2	3.9
	130821	16.8	62.5
	165286	0.7	6.8
Rock chips	158059	1.3	11.4
	DLP655	8.0	18.8
	UHO159	161.0	435.0
	SJH061	0.1	10.7
	SJH063	31.2	105.0
	SJH064	0.8	370.0
	GD0008	0.1	160.0

TABLE 2.4—Some combinations of parameters required to define drill targets for unconformity-related uranium deposits.

Parameter	Possible drill targets			
	Case 1	Case 2	Case 3	Case 4
Sandstone cover	X	X	X	X
Structure		X	X	X
Graphite	X			
Radioactivity			X	
Favorable host rock				X
Uranium anomaly		X		
Base metal anomaly				
Radon anomaly				
Favorable alteration				

X = essential criteria

deposit in the Athabasca Basin of Saskatchewan. For the period 1980 to 1985 industry has shifted almost entirely to exploration for Au, with discovery of Hemlo as one of the most noteworthy achievements (gross value in 1985 $7.5 billion).

In the southwestern United States exploration has focussed on locating epithermal gold mineralization, and guidelines for exploration for these types of deposits have evolved over time. Several possible models have been summarized by Silberman and Berger (1985). These were given the following descriptive labels:

(1) Low sulfur, quartz-adularia, Bonanza IA model.
(2) High sulfur, quartz-alunite, Bonanza IB model.
(3) Low sulfur, quartz-adularia, Hot Springs IIA model.
(4) High sulfur, advanced argillic-alunite, Hot Springs IIB model.

For purposes of discussion the Bonanza IA model has been selected to provide a focus for a series of questions which are designed to translate the experience gained in studying the unconformity U model into the design of an exploration program for this particular type of epithermal gold deposit. Definite answers to most of the questions are not possible and a range of responses is reasonable. Representative answers are provided at the end of the chapter.

1. a. What are the basic empirical geological properties of the Bonanza epithermal Au deposit type?
   b. Propose a geochemical model which could be used to guide exploration for this deposit type.
   c. What elements would you select as pathfinders in the search for these deposits? Would you want to determine concentrations of elements other than pathfinders; if yes, which elements and for what purpose(s)?
   d. Are there any peculiarities in potential pathfinder elements which might be useful?
   e. Are there any geochemical problems which might be encountered and which have to be overcome to successfully use geochemical methods?
   f. From what you know of the opinions of company geologists, geophysicists and geochemists, prepare a table, analogous to Table 2.4 for U, which summarizes what could be considered as drill targets for the company.
   g. Would you select total or partial digestion procedures for Au and/or the pathfinder elements? Why?

PHASE 2—THE FIELD ORIENTATION VISIT

The unconformity-related uranium deposit

Introduction

Orientation studies are conducted to determine optimum sampling and analytical techniques and criteria for the interpretation of subsequent survey data. They are best conducted around known mineral prospects, but in the absence of known occurrences (or if access to suitable properties is prevented by hostile landholders) studies are conducted in representative environments to determine guidelines for subsequent sampling. Procedures for orientation studies have been outlined in Chapter 1.

Figure 2.4 depicts prospective areas for unconformity related deposits in North America. All lie in Canada. The peculiarities of each will be described in turn.

Athasbasca Basin

The Athabasca Basin of northern Saskatchewan has seen the greatest intensity of exploration activity. Basin geology is dominated by sandstones of the Athabasca Formation. The landscape is flat with coniferous forests and numerous lakes. Soils are derived directly from bedrock or, more commonly, from sand-rich glacial till and fluvioglacial deposits. The coarse texture of these materials has promoted excellent soil development in which a typical profile has the following characteristics (soil horizon nomenclature is explained in Chapter 3):

(1) A surface organic horizon (LH) of leaf litter and decomposed humus—1 to 10 cm thick.
(2) An extremely leached mineral horizon (AE), immediately below the LH horizon, having an irregular lower boundary. Average thickness of the AE is 50 cm.
(3) A rusty, red brown, iron oxide-rich horizon (BF) 20–50 cm thick. The rusty colors are replaced by various shades of brown below 100 cm depth.

The profile description suggests problems are likely if routine soil sampling methods are used. Soils are very sandy, and it is to be expected that large samples would have to be taken to gather sufficient minus 80 mesh material. This would take time and, even if collection was successful, the minus 80 mesh fraction would probably consist predominantly of a quartz-feldspar sand. This cannot easily retain a metal as labile as U. To avoid the leached horizon, sample depth will likely be close to 1 m and sampling will be an expensive procedure.

Should soil sampling be considered?

Soil or deep overburden U and pathfinder element anomalies have been identified at most of the U deposits, but workers have used the minus 200 mesh fraction. Based on the description of the soil profile and on chemistry of U, an alkaline reducing extraction specific for dissolving the iron oxide fraction might be suggested instead of a routine digestion. This would have to be confirmed by orientation studies. Use of humus samples and biogeochemistry are other approaches. All these procedures tend to emphasize U derived by hydromorphic dispersion from its bedrock source.

At the Cluff Lake deposit, overburden comprises till and fluvioglacial outwash deposits. These have the same general soil profiles described above but are telescoped over a total thickness of 0.5 m. Till deposits have a sand-silt-clay texture and contain numerous stones, including the abundant radioactive boulders that were responsible for the radiometric discovery of the deposit. Glacial outwash, dominated by stratified boulder layers, is more sandy. Highest U and Pb values are found below the soil A and B horizons in the parent material C horizon.

A routine soil survey is appropriate for these environments, recognizing that it is focussing on the glacial dispersion trains rather than their bedrock sources. However, the latter might also be outlined under favorable conditions. This is a benefit rather than a burden to initial exploration because these glacial dispersion trains are much larger exploration targets than their source in mineralized bedrock. At Cluff Lake the radioactive boulder train outlines the same anomaly as the soil survey, but in other areas this might not be the case. Followup exploration would attempt to trace the glacial dispersion train towards the root zone of the anomaly following standard geochemical models described in Chapter 6.

The Athabasca Basin is relatively well drained. Local topography will have some effect on geochemical distributions but overall will not be much of a factor. Coarse texture of the soil should promote experimentation with radon gas techniques. Abundant lakes, acting as sinks for hydromorphically transported metal, should provide a good reconnaissance sampling medium. Outcrop exposure is generally limited and geological mapping will probably be of limited value.

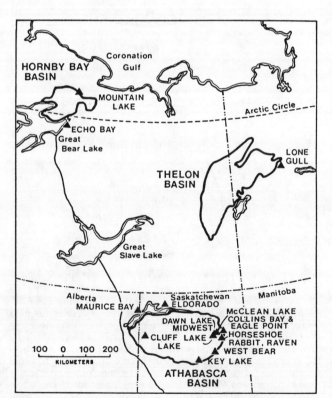

FIGURE 2.4—Helikian basins of the northwestern Canadian Shield. After Hoffman (1983).

Thelon Basin

The Thelon Basin in the eastern Northwest Territories is in arctic Canada close to Hudson's Bay. The region is north of the treeline and the landscape is generally flat. In the east it is dominated by old strand lines (gravel beach deposits) of Hudson's Bay. Outcrops are present on hills but large expanses of ground are covered by overburden consisting predominantly of glacial till with intercalated fluvioglacial outwash deposits. Glacial deposits are generally less than 10 m thick although locally they may exceed 30 m.

The area is within the permafrost zone and overburden is characterized by development of frost boils. These are upwellings of material from 1 m or greater depths as shown in Figure 2.5. Two types of soil material are therefore available: normal soils, with a cover of grasses and scrub brush, and the frost boils without covering vegetation.

Soils are poorly developed except in sandy, well drained areas on top of bedrock or outwash flats. Profiles typically consist of a layer of organic accumulation (LH), a brown sandy layer apparently leached of clay minerals (AE) and either a brown zone of weak clay enrichment (BT) or a brown to gray mottled zone (BG). Mottles may not be present in some areas (C1). Soils are commonly water saturated by melting of permafrost in the talik (thawed zone) and, depending on time of year, permafrost is often within 1 m of the surface. Clay layers are formed at the permafrost boundary. The ground is easily liquified during sampling and in some cases soil profiles cannot be dug without pit wall collapse. Iron-rich (BF or BM) and white leached (AE) horizons are rare and restricted to well drained environments associated with beaches, terraces and outcrop areas.

Frost boil colors are sometimes diagnostic of underlying bedrock and can be used to assist geologic mapping. They can exhibit easy liquifaction but when dry their caps become very difficult to penetrate with a shovel. A white deposit formed on the surface of some dry caps has a pH of about 10.

The landscape is characterized by numerous lakes and widely spaced streams. Drainage channelways are often overgrown by bog vegetation which is always frozen below depths of 30 cm. This must be considered in soil surveys because collection of the underlying inorganic material will be difficult if not impossible. Streams rarely have sediment and reconnaissance surveys would have to rely on lake sediment geochemistry.

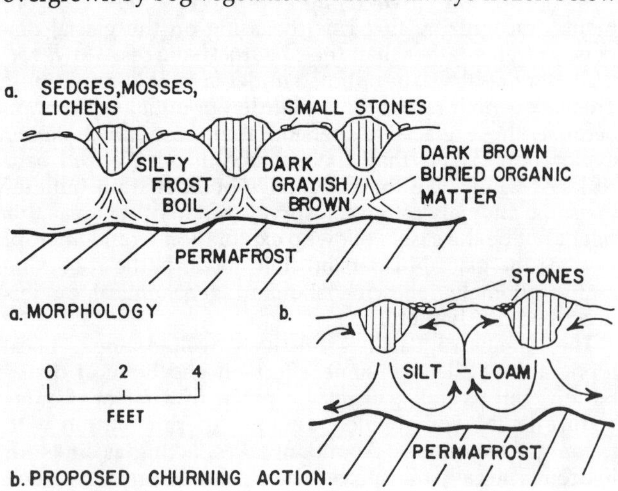

Exact controls exerted by permafrost on trace element dispersion are incompletely known. It might be predicted that solution of metals from mineral grains in the overburden and their hydromorphic dispersion in the water saturated talik would be a more active process than metal dispersion in thin layers of water (microlayers) surrounding mineral grains in frozen ground. Thickness of the talik would then possibly be a factor of geochemical significance. Figure 2.6 illustrates that permafrost is at a shallow depth beneath bogs and thawing of clay-rich overburden is more limited than sandy overburden. Beneath lakes the talik is particularly deep. A bedrock source of U lying at a shallow depth beneath or beside a lake might be expected to develop related geochemical anomalies in soil, whereas the same prospect sufficiently removed from a lake would not produce the same type of anomaly. Sample texture and proximity to lakes are therefore likely to be important factors in interpreting geochemical distributions in permafrost terrains.

Hornby Bay Basin

The Hornby Bay Basin exhibits many similarities with the Thelon Basin, except relief is much more extreme, outcrop

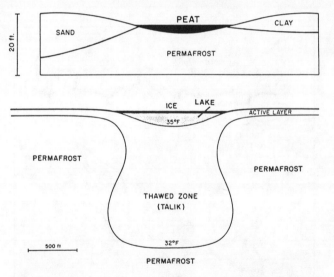

FIGURE 2.5—Idealized cross sections of silty frost boils. After Levinson (1980).

FIGURE 2.6—A: Diagram illustrating differences in depth to permafrost in the continuous zones, and in those parts of the discontinuous zone underlain by permafrost, resulting from differences in types of surface materials. Because peat beds are good insulators, particularly when they are dry, permafrost is closest to the surface beneath them, and the active layer in them is usually thin. The active layer is thickest in well-drained, sandy soil, where the ground freezes and thaws more deeply than in clay. B: Cross section of an ice-covered lake and thawed zone in the permafrost region in northern Canada. The ice is four feet thick; the active layer is two feet thick; and the lake is 10 feet deep (scale distorted in diagram). Bottom water temperature in the lake is 35°F. After Levinson (1980).

more abundant and areas of overburden cover more restricted. The arctic treeline crosses the southern limit of the Basin. Features in common with the Thelon Basin include climate, permafrost, frost boils, water saturation of the talik, limited soil development, poor surface drainage and abundant lakes.

Outcrops in permafrost environments tend to be affected by frost heave. Boulder fields, comprising large angular blocks of rock, are believed to have been heaved from bedrock lying beneath 1 or 2 m of overburden. Mechanical transfer of metal to the fine fraction during this process is to be expected and a local geochemical signature is likely to be imparted to the overburden regardless of its original genesis. A residual geochemical model is therefore probably at least partly applicable to the interpretation of results associated with these boulder fields.

Other parts of the landscape have a characteristic striped appearance resulting from solifluction lobes moving downslopes. Geochemical planning and interpretation will have to consider this factor in selecting grid orientation, and in predicting dispersion pathways and the sources of metal in anomalous areas.

Fifteen orientation studies were conducted around known U occurrences at Hornby Bay to determine the mechanism and extent of geochemical dispersion of pathfinder elements as well as U. Such information is critical to establishing sample density criteria. Table 2.5 summarizes lengths of dispersion trains, maximum U concentrations and pathfinder element associations. Strong correlation of U and Cu is indicated. Cold extractable total heavy metals (THM) determined by the Bloom test (Bloom, 1955) or the copper specific Holman test (Holman, 1963) could therefore be used to locate U occurrences. Pb is also associated with the U occurrences. However, because of its relative immobility, Pb accumulation in soils is restricted to rubble overlying U occurrences.

The preliminary field visit—is it necessary?

Two options are available to the planners of a geochemical survey:

(1) Select off-the-shelf, proven technologies.
(2) Question the merits and possible pitfalls in using standard technologies within the area of interest.

In the first case planning is straightforward and the geochemical survey can be conducted without any preliminary fuss and bother. Most important, planning overhead costs are at a minimum. This option can probably be accomplished without a visit to the field, but the explorationist should at least consult published topographic maps and air photographs to ensure objectives can be met by the planned program. However, the probability that this option will ultimately prove satisfactory in exploration effectiveness and efficiency is low. For example:

(1) In the Athabasca Basin would standard soil sampling and a standard analytical digestion be appropriate?
(2) In the Thelon or Hornby Bay Basins, would soil and/or frost boil sampling be applicable?
(3) Do landscape conditions affect geochemical exploration strategy?

In answering these questions it is apparent that serious difficulties with off-the-shelf techniques could adversely affect the outcome of an exploration effort. Severity of the problem might vary from a minor setback to the total failure of an expensive exploration program. The cost to obviate these concerns will vary from area to area but will probably be minimal compared to the large sums of money at risk. Orientation studies and expert advice should not be circumvented.

TABLE 2.5—Dispersion train length, maximum uranium values and pathfinder element associations from soil orientation studies, Hornby Bay Basin.

Prospect name	Dispersion train length for U (m)	Maximum U value (ppm)	Base metal associations
MAC	20	6.6	Pb, Cu
WOLF SOUTH	15	6	Pb, Cu
WOLF NORTH	120	13.5	Pb, Cu
SOUTH	10	4	Cu
BEAR reconnaissance phase	120	30	Cu
BEAR detailed evaluation	200	124	Cu
G	20	1	
BESS	40	10	Pb, Cu, Co
PARC	10	11.8	Pb, Cu
WOLF EAST	95	4.0	Cu
FLOW SOUTH	115	5.0	Pb
ECHO	110	22.0	Pb, Cu, Ag
CAM	160	11.0	Pb, Cu, Ag
MUNCH	145	15.8	Pb, Cu, Ag
TABB	180	455	Pb, Cu
DDR	25	4.6	Ra*, Po*
NWS	200	29.0	Ra*, Po*, Cu, Co, Ni

*Ra and Po determined only at DDR and NWS prospects.

The epithermal gold deposit

2. a. List at least four factors which might be important to conducting a geochemical survey. Describe the procedures you would use to identify potential problems with each.
   b. Can any peculiarities of the landscape be used to advantage in the geochemical survey?

CONTINUED OFFICE PLANNING

The unconformity-related uranium deposit

If field orientation studies had been conducted, sample spacing, soil horizon effects, optimum size fractions, digestion procedures and soil and overburden pH and Eh conditions would be evaluated in addition to the influence of landscape, overburden composition, drainage conditions, abundance of outcrops, glacial effects (if any), etc. For example, studies in the Athabasca Basin suggest routine soil sampling would be appropriate for some areas but not others. Its selective application releases funds for other methods to explore prospective areas where use of soils is inappropriate. Table 2.6 summarizes principal exploration and geochemical parameters to be considered before initiating a geochemical survey.

> Would a frost boil or B horizon soil survey be appropriate for U exploration in the Northwest Territories?

Published studies on Cu exploration in the late 1960's recommended frost boil sampling, rather than B horizon soils, on the premise that frost boils represented material from deeper in the profile and were not as depleted in metals by leaching. They should, therefore, provide better anomaly contrast. It might seem reasonable to extend these arguments to U and perhaps even reanalyze archival samples. However, the frost boils have neutral to alkaline pHs and under these conditions U is very mobile and leaches rapidly. The leached U is transferred to the adjacent soils which are weakly to moderately acidic as a result of decay of surface vegetation. A B horizon soil survey is therefore to be preferred for U.

In addition to the geochemical survey, a project manager must consider the conduct of the entire exploration program. Specific planning for geological, geophysical and prospecting work is beyond the scope of this volume. The geochemical survey will, however, require maps showing geology, geophysics, mineral occurrences, topography, cultural features and land status, all at the same scale. Airphoto coverage at any available scale is also needed.

Field sample locations are plotted on base maps showing topography. These should be updated regularly to show positions of mineral occurrences, access roads and trails. Recommended field scales are 1:5000 (1 cm = 50 m); 1:4800 (1" = 400 ft), 1:6000 (1" = 500 ft), 1:10,000 (1 cm = 100 m) and 1:12,000 (1" = 1000 ft). Scales larger than 1:5000 should be considered only for orientation studies or where the sample interval is closer than 25 m. Serious consideration should also be given to ensuring a degree of permanency to soil grids by constructing survey monuments and using aluminum tags for identifying sample sites.

Manpower requirements for the soil survey must also be established. This involves estimating the number of soil samples to be collected and sample density. Difficulties in sampling must be anticipated so that the number of samples per man day can be estimated and an appropriate number of samplers retained for the project. It is suggested that serious consideration be given to employing a qualified sampling supervisor and an interested, intelligent group of sample collectors.

The epithermal gold deposit

3. a. Can you suggest any features in the geochemistry of Au and its pathfinders that should be addressed in an orientation sur-

TABLE 2.6—Summary of the principal exploration and geochemical parameters affecting the decision to conduct a geochemical survey and the type of survey to be undertaken.

Parameters	Athabasca	Thelon	Hornby Bay
1. Number of ore reserves	12	1	2
2. Number of mines	4	0	1
3. Exploration activity	High	Low	Low
Overburden type	Glacial	Glacial	Glacial
Overburden transport	Extensive	Local/extensive	Local/extensive
Climate	Boreal forest	Permafrost barrens	Permafrost barrens
Topography	Flat	Flat	Flat/rugged
Drainage pattern	Lakes	Lakes	Lakes
Soil type	Podzols	Regosols	Regosols
Soil texture	Sandy	Sand-silt-clay	Sand-silt-clay
Bedrock outcrop	Poor	Poor	Excellent
Overburden thickness	Thick	Thick	Thin
Soil drainage	Excellent	Poor	Poor
Boulder trains	Present	Present	Present
Geochemical peculiarities	Few fines	Frost boils Old beach strand lines	Frost boils

vey? What procedures might you use in such a study?
b. If you have identified a potential problem with analysis as one of (a) above, how would you resolve this dilemma?
4. Assuming $100,000 is available for soil geochemical studies, prepare a work program for a survey, based on a 50 × 100 m sample grid, in an area you are familiar with. Estimate the number of samples that might be collected if a moderate degree of difficulty is encountered. Your estimate will have to consider analytical and reporting costs and recognize normal field-related problems, such as down-days. What factors have you considered in arriving at your answer?

REGIONAL EXPLORATION

The unconformity-related uranium deposit

Soil surveys are not normally used for regional exploration unless landscape conditions favor their application. In all three Basins a standard lake sediment survey would probably be selected for reconnaissance geochemistry. Can reconnaissance soil sampling be used in any of the three Basins?

Based on a description of the landscape as having significant topographic relief and a striped appearance (due to solifluction), reconnaissance soil traverses can be recommended at Hornby Bay. It is presumed that available maps or concurrent mapping can establish the approximate position of the unconformity. Once this is established, active downslope mechanical and hydromorphic dispersion should permit use of a traverse line, strategically positioned along the base of slope, to assess the U potential of the unconformity zone. Figure 2.7 illustrates different geological and landscape conditions together with the suggested location of the proposed traverse line. The same procedure cannot be used in the Athabasca or Thelon Basins because of their subdued relief.

Assessment of potential targets selected by other exploration methods, such as geophysics, can also be undertaken using reconnaissance soil sampling. If a U-rich structure can be differentiated from a U-poor structure, then a reconnaissance soil survey has the potential of rating geophysical targets for order of drill testing. Sampling would be undertaken either downslope of the suspected bedrock source,

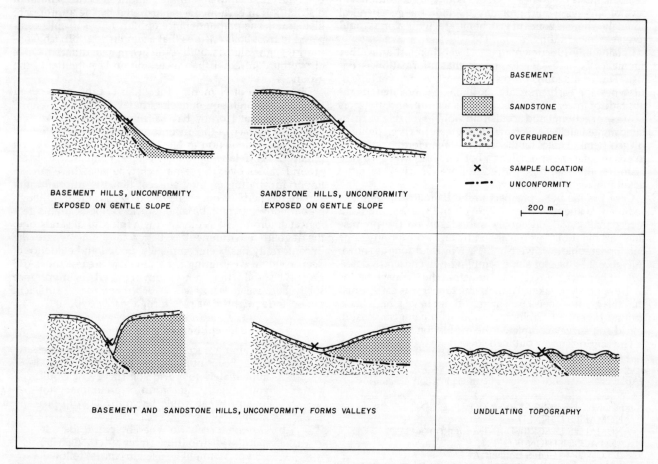

FIGURE 2.7—Relationship between reconnaissance soil sampling site and topographic position of the unconformity zone. After Hoffman (1983).

to intersect hydromorphic dispersion trains, or down ice to cut glacial dispersion trains.

The epithermal gold deposit

5. a. Can reconnaissance soil sampling be used in your exploration program? What geochemical model is guiding your selection of this reconnaissance procedure?
   b. Is it likely that you will encounter two or more different geological or other types of geochemical environment in your work? What could these be? Provide an example illustrating the method you would use to focus your attention on the most prospective ground.

THE ROUTINE SOIL SURVEY

The unconformity-related uranium deposit

Table 2.7 presents a typical exploration scenario once the reconnaissance survey is completed. Geochemical anomalies have to be rated relative to each other and with respect to geophysical and radiometric anomalies and features of geologic interest. Coincidence of anomalous conditions on several surveys would probably upgrade the geochemical anomaly whereas zones of geochemical activity alone would receive a lower rating.

Cataloguing (documentation) of geochemical and other anomalies is necessary to ensure that information is not lost. The cataloguing process summarizes information of all types for each anomaly, outlining factors that might upgrade or lower the significance of a feature, and indicates a course of action. Land acquisition would probably be instituted immediately if a worthwhile anomaly has been defined. All geochemical notes for the area must therefore be reviewed to ensure that the anomaly is a bona fide feature. Two case histories illustrate this: procedures are described in more detail in Chapters 3 and 6.

Case 1 is the BOG prospect found by followup of a lake sediment U anomaly (Figure 2.8A). Land was acquired and a reconnaissance soil survey undertaken on the premise that basement hills were flanked by an unconformity zone and topographic lows were underlain by sandstone as shown in Figure 2.7. Base of slope sampling was chosen to determine if bedrock sources of U lay within the drainage basins of the anomalous lakes. At the same time it was established that the granite gneiss basement did not have a high background U content. Preliminary followup being successful, a grid soil survey completed the evaluation.

The routine survey outlined several areas of interest with an anomaly threshold of 1.3 ppm being exceeded by about 5% of the data (Figure 2.8B). Higher contrast anomalies outlined to the north subsequently led to the discovery of several U occurrences. One of these, the BOG prospect, represents a significant discovery where a stockwork of pitchblende veins has been delineated over an area of 125 X 200 m. Radioactive boulders, similar to those at BOG and grading up to 3% uranium, have been found 3 km further north and probably represent a glacial boulder train.

The second example illustrates a portion of a regional soil survey (Figure 2.9A). Subsequent mapping and prospecting located several pitchblende veins in graphitic and pyritic zones in gneissic basement. Multisample U anomalies in the followup soil survey closely reflect the known radioactive zones (Figure 2.9B).

At Hornby Bay soil sampling is relatively straightforward providing that soils and frost boil materials are not indiscriminately mixed. The latter are more easily collected but are an inferior medium for U exploration. Figure 2.10 presents two interpretations of a survey where this factor was not considered important. The first (Figure 2.10A) shows how high U background levels in soils minimize the influence of weakly anomalous U conent in frost boils. The second (Figure 10B) shows how anomalies can "expand" if frost boil data are used to determine contour intervals. Uranium distribution in each media is illustrated by the appropriate histogram. The preponderance of high values in soils compared to frost boils ensures that a simplistic approach (contouring high values) would result in a map of different types of overburden rather than focussing on U potential of the ground.

Distribution of elements other than U can also be used to assist interpretation of geological favorability of an area. For example, at Hornby Bay basement rocks immediately below the surface of the unconformity are characterized by their relative depletion in Ni. Thus, at the TABB prospect (Figure 2.11), relatively low Ni values, compared to background values over the same rock type elsewhere, are an important finding as they suggest that the unconformity surface has perhaps not been eroded by glaciation. Although regolith developed on basement rocks is not as good a geological environment as one having a thin cap of sandstone, the regolith environment is better than one where the unconformity has been completely eroded and evidence of deep residual weathering is absent. The area of the TABB prospect should thus not be downgraded in importance solely because it is located in basement rocks: sufficient pitchblende might remain to form a viable deposit.

The epithermal gold deposit

6. a. Assuming reconnaissance gold anomalies have been outlined, would you proceed immediately to a grid soil survey or would you suggest some form of preliminary followup? What would be your followup program?
   b. Outline a scenario whereby pathfinder element(s) distributions rather than distribution of Au might be used to control followup proposals for the epithermal model.
   c. Equivocal results provided by radon geo-

TABLE 2.7—Sequence of events in an exploration program.

1. RECONNAISSANCE SURVEY COMPLETED
2. ANOMALIES DEFINED
3. ANOMALIES DESCRIBED AND PRIORITY RATED
4. LAND ACQUISITION IF OPEN
5. FOLLOWUP STUDIES UNDERTAKEN
6. ONGONG ASSESSMENT OF ANOMALIES IN THE LIGHT OF RECENT WORK

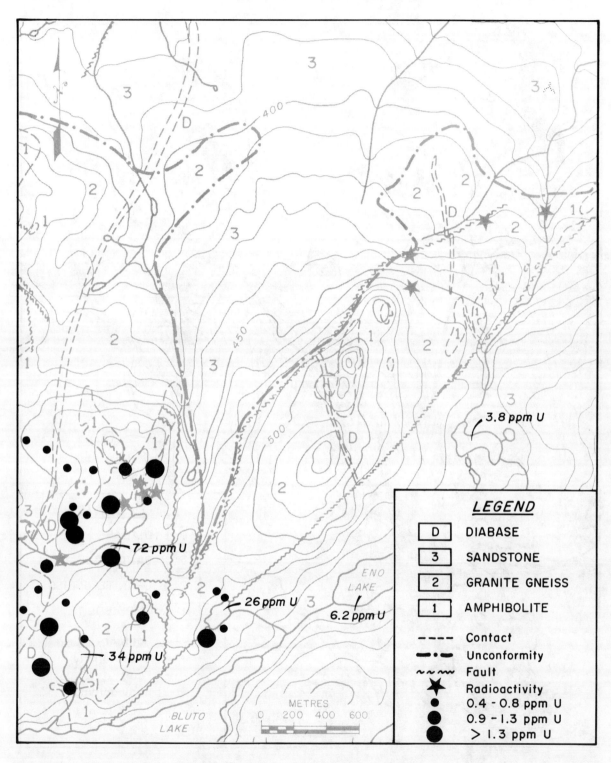

FIGURE 2.8—BOG U Prospect. A: Soil survey followup of lake sediment U anomaly.

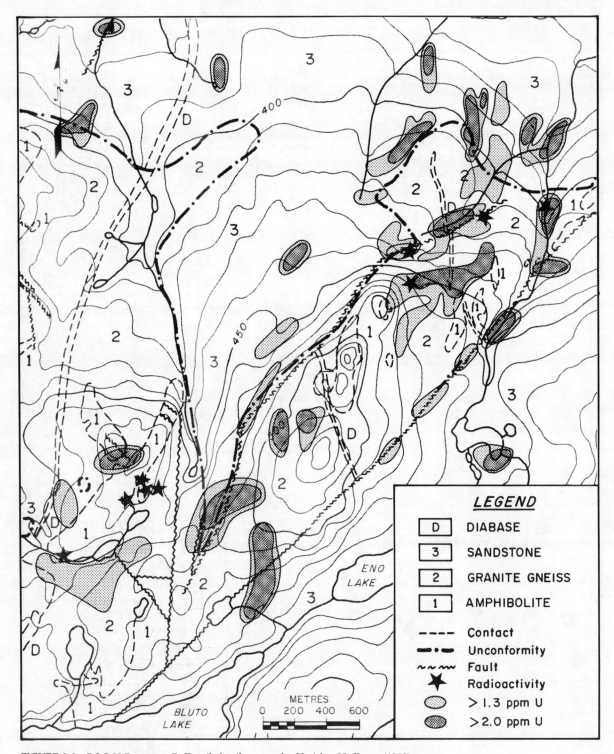

FIGURE 2.8—BOG U Prospect. B: Detailed soil survey for U. After Hoffman (1983).

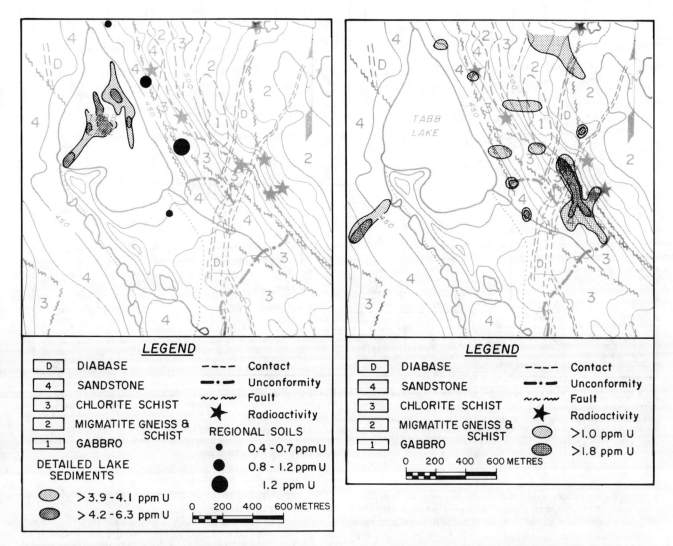

FIGURE 2.9—TABB U prospect. A: Regional soil survey for U. B: Detailed soil survey for U. After Hoffman (1983).

chemistry in the search for U have complicated its effective application (see Table 2.2). Is the same possible for Hg? Prepare a chart similar to Table 2.2 for Hg.
d. How would you rate your geochemical anomalies? What rating would geophysical anomalies and geological factors such as structure, alteration, etc., play in your decisions?
e. What followup methods would you recommend to test your best anomalies?

THE ULTIMATE TEST—THE DIAMOND DRILL PROGRAM

The unconformity-related uranium deposit

Diamond drilling is the ultimate evaluation tool for determining property potential. Preliminary surveys, such as a soil program, should be made as efficient as possible so that drilling, an expensive procedure, can be directed towards targets of real potential. Minimizing costs prior to the drill stage is a common management strategy. Defining drill targets rapidly is another objective—even if this is not explicitly mentioned by management. How often has the drill arrived on site before targets have been fully established or prioritized?

Cost saving for geochemical programs invariably begins early: low priced personnel, little or no instruction or day-to-day guidance, no basemaps, no permanent grid line markings and speed sampling. The cheapest analytical procedure is often selected rather than a more suitable technique or multielement analysis, even though these may ultimately be more effective. With poor geochemical work the stage is set for survey failure. A multitude of false anomalies are likely to be defined and insufficient information will be available to interpret geochemical data beyond con-

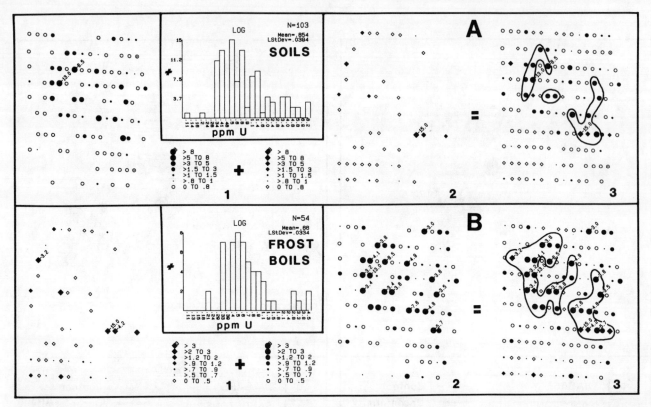

FIGURE 2.10—Two types of geochemical sample have been collected from the grid: soil (circles) and frost boils (diamonds). A size coded symbol plot of soil results is shown to the left of the histogram in (A); immediately to the right of the histogram frost boil results are shown size coded using the concentration intervals used for soils; and, the combined results (i.e., assuming that the sampler was unable to distinguish the two sample types) are shown at the extreme right of (A). The same procedure has been used in (B) to show (to the left of the histogram) frost boil results correctly size coded; soil results coded using the same concentration intervals as the frost boils; and (at the extreme right of (B)) the combined results. Note the differences between the patterns.

touring big numbers. The expensive drill program will probably test at least some anomalies that should have been dismissed in short order. Worse still, subtle geochemical features related to mineralization will have been missed!! Is this the way to run an exploration program?

> Based on the examples of Figures 2.8, 2.9 and 2.10 what rating do you think a 25 ppm U soil anomaly would merit? What if that anomaly represented a bog? What if personel on site did not recognize the problem?

Experience has shown that unless the company was fortunate, followup would often continue to the conclusion of an ineffective drill program.

Table 2.4 indicates that a soil U anomaly could provide the focus of a drill test if it lies in proximity to a structure or radioactive occurrence. Geochemical models (Chapter 6) enable prediction of probable source areas in bedrock concealed by glacial overburden. This is most straightforward when lodgment till is sampled and bedrock sources of metal can be predicted within a few hundred meters or closer. Often geophysical anomalies or other criteria are used to site the drill up-ice at or close to the predicted termination of the geochemical anomaly, but in many cases geochemical data alone might suffice.

Grid soil anomalies normally require additional followup before the drill can be sited. Remember the soil anomaly is the landscape expression of a three dimensional dispersion train that has its roots at the sought-after mineral occurrence. Trenching or deep sampling over the best soil anomaly will probably intersect nothing at depth unless the environment is one of truly residual soils. More usually trenches, pits or deep sampling must extend up ice and/or upslope, into areas of background in near-surface soils, to trace the metal to its bedrock source. These methods of followup might not, however, be needed if overburden cover is relatively thin and prediction of bedrock sources considered relatively reliable.

Caution must be exercised to avoid assuming that the nearest geophysical anomaly or mineralized boulder is the target if geochemical interpretation suggests otherwise. The former may represent targets in their own right, but too often a negative test of the closest geophysical feature is taken to indicate that the geochemical anomaly is some artifact of the surface environment and thereby merits no further attention. This could be a mistake and the explorationist should not be too eager to explain away a geo-

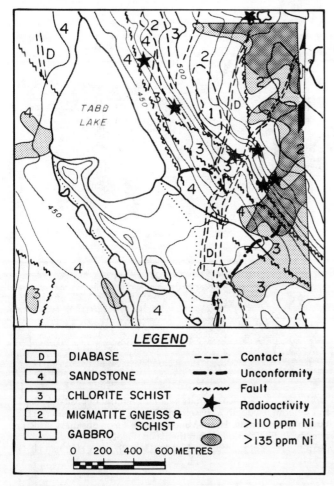

FIGURE 2.11—TABB U prospect—detailed soil survey results for Ni. After Hoffman (1983).

chemical feature until it is confidently traced to its source in bedrock.

Drill testing strategy for unconformity related U deposits at Hornby Bay is illustrated in Figure 2.12. The following principles play a prominent role in determining methodology.

(1) An unconformity-related U deposit is the target.
(2) Sufficient local material has been incorporated into the overburden to consider geochemical patterns in the soils to be partially residual. Alternatively, groundwater dispersion and solifluction have transported anomalies downslope up to 200 m (from orientation study results).
(3) Case 2 (Table 2.4) represents the synthesis of information leading to drill target selection.
(4) Diamond drilling is undertaken in virgin areas only if the sandstone is less than 150 to 200 m thick. Below this depth development and min-

ing costs will limit the economic potential of any mineralization encountered.
(5) The limited dimensions of the unconformity-related U deposit type (Figure 2.1) require fences of at least 3 diamond drill holes, rather than single holes, to evaluate an anomaly.

Procedures for drill testing at the margin of the Hornby Bay Basin (Figure 2.12A) or at a central position in the Basin (Figure 2.12B) are described in the figure captions.

The epithermal gold deposit

7. a. Summarize the key considerations for selecting a drill target.
   b. Provide a graphical illustration, comparable to Figure 2.12, to illustrate positioning of drill holes. The caption to this illustration should describe your thought processes.

CONCLUDING SUMMARY

The example of exploration for unconformity-related U deposits, augmented by questions regarding the same concepts applied to exploration for epithermal Au deposits, should indicate that the process is not as simple as one might initially think. A multidisciplinary approach to problem solving stands the best chance of leading to a successful outcome. The failure of many geochemical programs more often than not reflects inexperience or naive assumptions rather than a genuine inability of geochemistry to locate mineral prospects. Table 2.8 summarizes the state of affairs found in many exploration programs, but which can often be avoided by use of common sense in the initial stages of the project.

ANSWERS TO EPITHERMAL GOLD DEPOSIT QUESTIONS

1. a. The empirical features of the Bonanza epithermal gold deposit type can be summarized as follows (based on Berger and Bethke, 1985):
   1. Gold deposits can occur in any rock type. They are found within an envelope of hydrothermal alteration characterized by presence of quartz veins, silica flooding and opaline silica.
   2. Normal, high angle faults are present.
   3. Brecciation is common.
   4. Deposits lie at or near the present or a paleo-land surface.
   5. Tertiary intrusions may be present nearby.
1. b. Geochemical models are illustrated by Figures 2.13 and 2.14.
1. c. Traditional pathfinder elements for Au in an epithermal system include As, Sb, Hg, and Tl above ore; Ag associated with ore; and base metals indicating deeper levels and probable erosion and removal of ore. Other approaches might center on highlighting

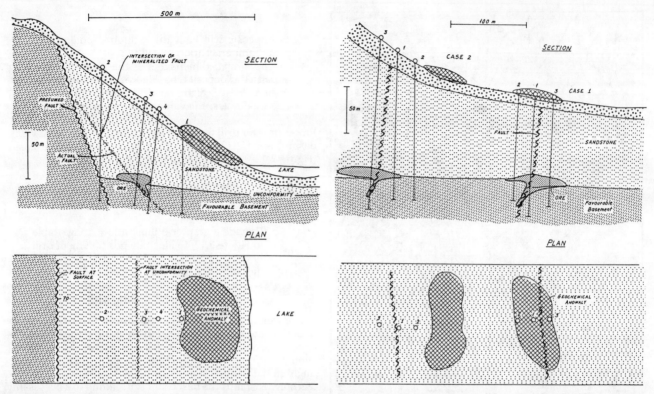

FIGURE 2.12—Procedure for drill testing of geochemical anomalies in the search for unconformity related uranium deposits. A: Exploration at the edge of the Basin. Hole 1 intersects the unconformity and determines depth to basement. Hole 2 attempts to intersect the host structure at the unconformity but either hits a fault contact or barren unconformity. Holes 3 and 4 search for a subsidiary structure at the unconformity. Note that no holes are sited on top of the soil anomaly. B: Exploration within the Basin. Depending on topography, drill holes will be sited on top of the soil anomaly (Case 1) or upslope of the zone of uranium-rich soils (Case 2). After Hoffman (1983).

TABLE 2.8—Denouement: six phases of a project.

1. WILD ENTHUSIASM
2. TOTAL CONFUSION
3. DISILLUSIONMENT
4. THE SEARCH FOR THE GUILTY
5. PUNISHMENT OF THE INNOCENT
6. REWARD OF NON-PARTICIPANTS

alteration effects such as locating a SO_4 cap above ore, clay (Al) alteration, adularia (K) or silicification.

1. d. The gaseous dispersion of Hg offers many of the same opportunities for detecting deeply buried ore zones as Rn geochemistry for U deposits. Many of the practical problems are also the same.
1. e. Orientation studies should be conducted prior to starting routine surveys. The search for epithermal Au deposits has centered on prospecting for Au-bearing sinter, vein quartz and silicified rock. It is possible that the Au might be very fine grained and perhaps encapsulated in silica. Weathering of such material might not release gold to the fine fraction (minus 80 mesh) normally used in exploration. Use of fire assay preconcentration on a coarse size fraction, rather than an aqua regia leach on the minus 80 mesh, might avoid the problem. Alternatively, samples might be pulverized to minus 200 mesh prior to analysis by either method.
1. f. Table 2.9 summarizes conditions which might be necesssary to define a drill target following the example of Table 2.4
1. g. A total digestion procedure would probably be selected for Au analysis to overcome the potential problem of silica encapsulation. Determination of elements such as As, Sb, Bi, Hg, Cu, Pb, Zn, Ag, etc., would be undertaken following a partial extraction to optimize identification of anomalous conditions. Cost of analysis for a partial extraction is also somewhat less than determination of absolute metal abundances.
2. a. Among the possible answers we offer the following as points for consideration:
   Soil horizon The type localities for the epithermal Au deposit-type in North America

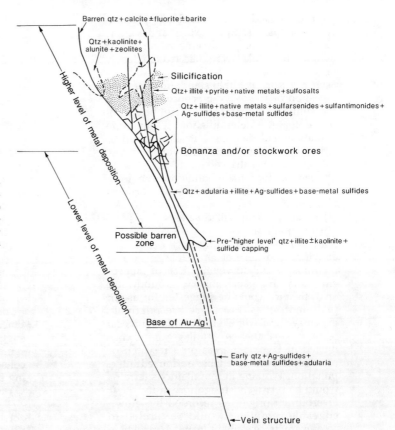

FIGURE 2.13—Schematic cross section of a quartz-adularia or low sulfur bonanza gold deposit, Bonanza-IA model, showing alteration mineralogy and two zones of mineralization from the "closed cell convection" model of Berger and Eimon (1983). After Silberman and Berger (1985). Reproduced from Reviews in Economic Geology, 1985, volume 2, chapter 9, p. 209.

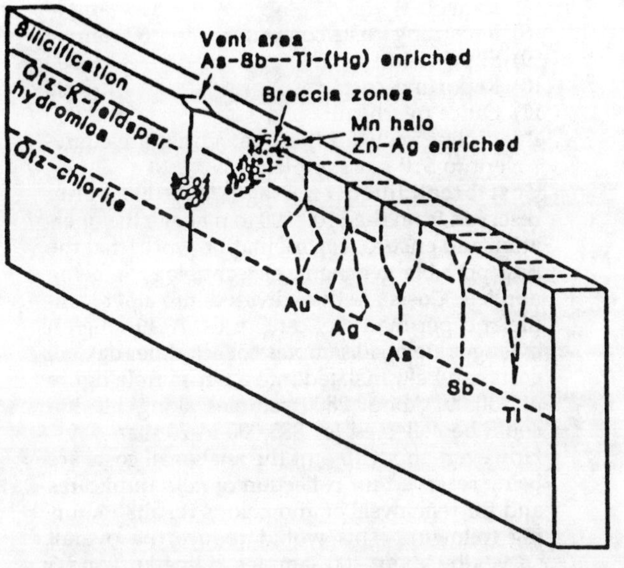

FIGURE 2.14—Schematic depiction of the spatial relationship of Au and Ag and selected trace elements to the generalized alteration zones in the hot-springs depositional model (from Berger and Eimon, 1983).

TABLE 2.9—Some combinations of parameters required to define drill targets for Bonanza type epithermal gold deposits.

	Possible drill targets			
Parameter	Case 1	Case 2	Case 3	Case 4
Siliceous sinter and/or silica flooding and/or quartz veins and/or opaline silica	X		X	X
Hydrothermal alteration	X	X		
Normal faulting		X	X	
Brecciation	X	X		
Tertiary landscape				
Tertiary intrusions				
Au occurrence		X		X
Gold geochemical anomalies	X		X	
Pathfinder element anomalies			X	
Geophysical anomalies				

X = essential criteria

have an arid to semi-arid climate. Soil development would likely exhibit a thin LH horizon, underlain by an AH and BM, followed by the CA zone. Orientation work would probably result in selection of the BM horizon. Quality of a survey could be verified by plotting of field notes describing soil horizon, soil color, sample texture and sample depth. Ca and/or pH determination would establish if samples were collected too deeply or too shallowly.

Sample preparation Either standard aqua regia digestion or fire assay preconcentration on the minus 80 mesh fraction might be appropriate for Au determination. Orientation studies would determine if silica encapsulation has resulted in retention of Au in size fractions coarser than 80 mesh.

Overburden Two main landscape environments characterize Nevada and northern California:

(1) broad, alluvium-filled valleys.
(2) mountain ranges.

Major valleys are filled by thick deposits of alluvium and are not amenable to normal methods of soil surveying. Use of biogeochemistry, Hg geochemistry, groundwater analysis, etc., could be suggested for use in pediment areas, but these are relatively expensive and will likely provide equivocal results. Nevertheless, studies can be conducted in favorable areas on a research or orientation basis to enable optimized interpretation. Working in mountainous terrains would be relatively straightforward.

Overburden origin, as recorded by knowledgeable samplers, would be plotted to accompany trace element maps. Areas where results are not likely to be overly informative could be identified.

Local conditions Survey planners should look for alternative sampling media to meet local landscape conditions. For regional work these include:

(1) Talus fines at or near base of talus slopes.
(2) Alluvial fan sampling at margins of large basins and valleys.
(3) Seepage zone sampling for pathfinders; sampling positions indicated by air photo interpretation.
(4) Playa lake sediment sampling.

Geochemical exploration procedures which are considered novel in approach require confirmation by means of an orientation study.

2. b. One aspect of the landscape that has useful attributes is topographic relief. Erosion since the formation of a mineral deposit may reveal deeper portions of the system. Geochemical patterns shown by the soil survey will then display evidence of the vertical zonation charateristic of this style of deposit and thus aid location of individual gold ore bodies.

3. a. The following are some of the points that would need consideration for soil surveys in western North America:

1. Application of the minus 80 mesh fraction and an aqua regia digestion could be problematical for Au. Pulverize a coarse fraction and/or test both the fire assay and aqua regia method for Au.
2. Sampling at a constant depth would produce spurious results for the pathfinder elements if the AH or CA horizons were inadvertently sampled instead of the BM. Collect soils from a constant horizon.
3. Spurious results might also accompany sampling of seepage zones at the base of major slopes. Identify this and other peculiar environments on field coding forms.

3. b. Laboratory staff follow their own routine procedures and it is unlikely they would recognize a problem such as silica encapsulation of Au within an individual batch of samples. Suggestions for optimizing sample procedures can only come from consultation between the field geologist/geochemist and senior laboratory staff. On occasion the laboratory will recognize a potential problem, but do not count on it.

4. Practical experience from surveys in Canada suggests that the following scenario is representative of a successful work program. Cost to collect soil samples is determined by the following factors:

1) Sampler salary/efficiency $2000/month/66%*
2) Logistical support $100/day*
3) Accommodation/meals $100/day*
4) Survey controls $100/km or $2/sample
5) Sampling materials $1/sample
6) Sampling speed   40/man day
7) Analytical costs (includes preparation and transport) $15/sample
8) Recording costs/computer $2/sample
9) Supervision $2500
10) Reporting costs $2500
11) Office overhead $10,000

*Sampling costs for these categories are equivalent to $10 per sample.

Cost breakdown envisages $15,000 being reserved from the $100,000 to manage the operation and ensure, via orientation work, that the best possible geochemical technology is being applied. Costs are then divided into a per sample and per man day categories. Assuming an average rate of 40 samples collected per day, all costs can be translated into a per sample figure of $30.00. About 2800 samples along 140 km could be collected for $85,000 in 70 man days. However, about 10% of the analytical costs are being reserved for collection of field duplicates and for reanalysis of anomalous results requiring followup. This would reduce the overall survey by about 100 samples (5 line km) and 3 man days.

Costs would increase by about $11 per sample in remote regions requiring helicopter assistance, assuming a $600/hour helicopter charge (fuel included), 3 hour minimum/day and four samplers carried by the helicopter. Only about 2100 samples (or 2000 after reserving funds for quality control) could be collected in this scenario and remain within budget.

5. a. Reconnaissance soil sampling can be used in many areas. Consideration of the models presented by Lovering and McCarthy (1978) suggests the following opportunities in the Great Basin of western North America:
   1. Talus fine sampling—residual model.
   2. Alluvial fan sampling—residual model.
   3. Drystream channels—residual model.
   4. Playa lake—residual model.

5. b. Having only two distinctive environments would be a highly unusual and favorable situation. Typically, each major geological unit would be reflected by its own suite of trace element levels. Geochemical environments represented by unique landscape conditions, overburden types, groundwater regimes, soil types, etc., would be subdivisions of each geological terrain. It is quite conceivable that you may recognize environments where routine geochemical sampling will not work and where it is unwise to pursue a geochemical survey.

   To focus on the most prospective ground, one must be able to discount spurious geochemical responses and follow genuine anomalies back to their bedrock source; see Chapter 6 on models and interpretation.

6. a. Reconnaissance soil survey results are likely to detect the first geochemical indications of a prospective area. These should first be confirmed by reanalysis of selected anomalous samples. At this stage it would be premature to devote a large proportion of exploration funds to a grid soil survey without some preliminary followup. Land would be acquired immediately in a competitive environment, but otherwise more reconnaissance work is needed to identify the best rather than the first targets in an area. Continued sampling of the same sort as the initial reconnaissance needs to be considered if the landscape permits this approach. Mapping to define the geological environments and prospecting to search for Au or evidence of an epithermal system (silicification, alteration) is necessary. A review of potential pathfinder element zonation patterns could also be informative. These studies would precede grid work, which would focus on those anomalies with the highest economic potential.

6. b. The epithermal gold deposit geochemical model has elements such as As, Sb, Hg and Tl surrounding and overlying the Au ore-bearing zone. Soil anomalies for these elements might indicate either lateral proximity to a Au-bearing zone or suggest Au will be found at depth beneath a pathfinder element anomaly. In the first case, the search for Au would be peripheral to the pathfinder element anomaly, in the second case a diamond drill hole would be needed to test for Au at depth.

6. c. See Table 2.10

6. d. Geochemical Au anomalies are undoubtedly rated relative to their magnitude and size, although there is no fundamental basis for this method of prioritization. Beyond these considerations, element zonation patterns of pathfinders associated with Au must be considered: Au anomalies accompanied by strong pathfinder element patterns (As, Hg, Tl, Sb) rating above Au alone; Au anomalies accompanied by strong enrichment of base metals would rate below Au anomalies alone on the premise that erosion has probably removed most of the Au zone, if one existed. Presence of favorable alteration comprising extensive silicification or clay alteration would be a positive factor. These last features might be amenable to mapping by geophysical sur-

TABLE 2.10—Factors influencing mercury distribution in the surficial environment.

Factor	Comments
Mercuric chloride and other soluble forms of Hg	Can be transported in groundwaters and form hydromorphic Hg anomalies at sites remote from the bedrock source
Mineralized float	Cinnabar can be transported as a heavy mineral and accumulated elsewhere under favorable conditions
Soil and overburden Organic content	Hg vapor and ionic Hg can be adsorbed at the base of the organic layer
Texture	More porous overburden will promote dispersion of Hg vapor
Bedrock structure	Hg emanations will concentrate along brecciated zones within faults giving a greater than normal flux of Hg to the soil

veys using resistivity measurements. Coincident geochemical anomalies with VLF-EM conductors would rate a high priority as being probably located over high angle structure controlling the mineralization.

6. e. If the current level of erosion is suspected of intersecting a commercial gold bearing zone lying beneath shallow overburden, trenching to expose bedrock is warranted. Remember that soil anomalies are likely to be transported downslope (or down ice if valley glaciation is an important parameter). Bedrock intersected by trenching should be continuously chip sampled at 3 to 5 m intervals or closer, sample collection involving the taking of about 1 kg of representative material per meter of trench. If overburden is beyond the limits of trenching or if you believe you are above the zone of gold mineralization, then drill testing is in order. Procedures for drill testing are reserved for the answer to question 7.b.

7. a. Au soil anomalies can be used to define drill targets (Table 2.9) but additional features add credibility to the target. Important criteria include the presence of sinter or silica flooding (silicification), and/or brecciation, and/or hydrothermal clay alteration. Structures are also needed to have focussed Au deposition.

7. b. Figure 2.15 provides an illustration of the philosophy.

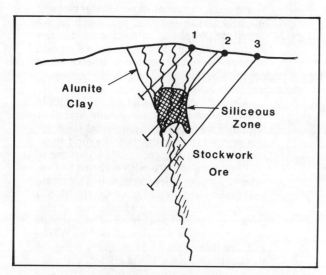

FIGURE 2.15—Drill testing of an epithermal Au prospect following the quartz-adularia/low sulfur Bonanza-IA model. A silicified fault zone is exposed at the surface. Drill hole 1 intersects the silicified fault. Grades and widths of Au in Hole 1 are insufficient to classify as ore but encourage further drilling following the mineralogical and/or geochemical zonation patterns of Figure 2.13. Hole 2 intersects the zone of silicification and Hole 3 locates the underlying stockwork ore.

REFERENCES

Berger, B.R. and Bethke, P.M. (editors) 1985. Geology and Geochemistry of Epithermal Systems. Reviews in Economic Geology Volume 2. Society of Economic Geologists, 298 pp.

Berger, B.R. and Eimon, P.I. 1983. Conceptual models of epithermal precious metal deposits. In: Shanks, W.C. III (editor), Cameron Volume on Unconventional Mineral Deposits. Society of Mining Engineers, p. 191–205.

Bloom, H. 1955. A field method for the determination of ammonium citrate soluble heavy metals in soils and alluvium. Economic Geology, v. 50, p. 533–541.

Bradshaw, P.M.D., Clews, D.R. and Walker, J.L. 1979. Exploration Geochemistry, Second Edition. Barringer Research Ltd., 53 pp.

Hoffman, S.J. 1983. Geochemical exploration for unconformity-type uranium deposits in permafrost terrain, Hornby Bay Basin, Northwest Territories, Canada. Journal Geochemical Exploration, v. 19, p. 11–32.

Holman, R.H.C. 1963. Field and Laboratory Methods used by the Geological Survey of Canada in Geochemical Surveys, No. 2; A Method for Determining Readily Soluble Copper in Soils and Alluvium. Geological Survey of Canada Paper 63-7, 5 pp.

Levinson, A.A. 1980. Introduction to Exploration Geochemistry, Second Edition. Applied Publishing Ltd., Wilmette, Illinois, 924 pp.

Lovering, T.G. and McCarthy, J.H. (editors) 1978. Conceptual models in Exploration Geochemistry—the Basin and Range Province of the western United States and northern Mexico. Journal Geochemical Exploration, v. 9, p. 113–276.

Rose, A.W., Hawkes, H.E. and Webb, J.S. 1980. Geochemistry in Mineral Exploration, Second Edition. Academic Press, 657 pp.

Silberman, M.L. and Berger, B.R. 1985. Relationship of trace element patterns to alteration and morphology in epithermal precious-metal deposits. In: Berger, B.R. and Bethke, P.M. (editors), Geology and Geochemistry of Epithermal Systems. Reviews in Economic Geology, Volume 2, p. 203–232.

Chapter 3

SOIL SAMPLING

S. J. Hoffman

INTRODUCTION

Samples collected on a geochemical soil survey can have a highly variable composition. This variability represents a major source of geochemical noise that cannot be entirely eliminated but can be minimized by prudent sample collection. To illustrate, consider the sampler confronted with a choice of black, white or reddish-brown material at the same site. Soil color is an easily observed property, related to soil composition, and it would appear intuitively obvious that it would be wise to collect material of the same color, if possible, at all sites on a soil grid.

Soil color and similar obvious differences in composition can be important indicators of variations in metal content. For example, organic and inorganic soils may react quite differently to an influx of metal derived from a mineral occurrence. Moreover, background levels of many metals in the two types of material could be expected to be markedly different. To ignore these differences ensures that some, and perhaps the majority of, exploration decisions could be misguided. Anomalies recommended for followup could be "oranges" in an "apple" data set rather than being "big apples" among otherwise ordinary "apples". The two most serious problems are:

(1) Significant anomalies are not outlined.
(2) False anomalies become the focus of exploration activity, and exploration funds are exhausted before bona fide features represented by (1) are recognized.

False anomalies are commonly related to unusually high levels of components of the soil sample, such as organic matter or clay minerals, scavenging trace elements contained in groundwater. Hydromorphic dispersion is arrested by scavenging agents much as a sponge can arrest a trickle of water flowing along a table top.

Failure to define anomalous conditions related to mineralization is also a concern. This can arise, for example, where samplers have failed to penetrate a leached zone near the surface, typically as a consequence of "speed" sampling. Such samples can be collected rapidly, but all evidence of mineralization may have been removed.

Borrowing an illustration from geophysics, consider a ground magnetometer survey conducted by a student unfamiliar with the importance of removing a pencil magnet from his vest pocket. At each station the magnetometer is held a variable distance from the magnet. If this were to become known, what project manager would accept the survey? Most would not hesitate to repeat the survey if magnetic readings are needed for the exploration program. The same philosophy should guide acceptance/rejection of results from a geochemical survey. Recognizing potential sources of error and avoiding them is the purpose of this chapter.

To illustrate the problem, consider a soil survey for Cu and Mo on a copper property in British Columbia (Figure 3.1). Anomaly thresholds of 50 ppm Cu and 5 ppm Mo were established.

> As project manager, what decision(s) would you make to initiate a program of followup?

Before you decide, here is some background information. The property, acquired as a result of a prominent lake sediment Mo anomaly, is underlain by a Topley intrusion—the same unit that hosts the Endako molybdenum mine 50 km to the west. Topography is relatively flat with lodgepole pine forest; lakes are numerous; stream drainage is well developed and bogs are distributed erratically over the landscape. Outcrops amount to 5% of the landscape and soils are generally thin, well drained and locally derived.

Returning to the question, what followup procedures would you recommend? Operators of the program decided that more detailed soil sampling would be an inexpensive way to focus attention on the highest contrast portion of the anomaly. Samplers recognized the occurrence of bogs in the area and were prepared to penetrate thick accumulations of organic matter with an auger. After five days of followup, samplers noted that almost all anomalies were associated with boggy areas. Comparison of the soil geochemical maps with an airphotograph of the property confirmed this. The followup program lasted an additional ten days, but analysis of the inorganic followup samples provided disappointing results. Cu and Mo values were much lower than initial survey data and claims were allowed to lapse.

> Reviewing the history and results, what instructions would you have given the soil samplers? What differences in procedures would you initiate to avoid the same pitfalls? Would you agree that had the

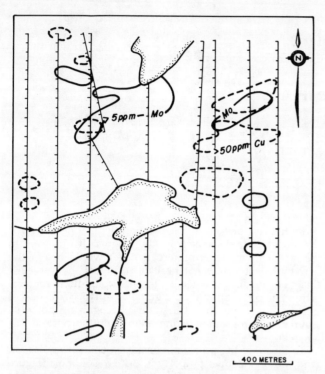

FIGURE 3.1—Geochemical survey for Cu and Mo, central British Columbia, Canada. The 5 ppm Mo and 50 ppm Cu contours outline distribution of boggy areas.

organic-rich nature of the anomalous samples been recognized before the followup program commenced, a different set of recommendations would have been issued? What would you have suggested after receiving results of the initial survey? The answer to these questions will become apparent after a discussion of the factors affecting the metal content of a soil sample.

To be effective, soil surveys have to avoid introduction of excessive noise from the sampling of varying types of materials. This is not an entirely achievable objective, but common sense during sample collection should reduce noise appreciably. An individual interested and trained in proper methods of sample collection is needed to conduct the sampling, balancing sampling speed with quality, to produce a cost effective exploration procedure.

The decision to undertake a thorough soil survey must be made by the exploration organization as sampling rates will likely fall from the 80 to 150 or more samples per man day of a "speed" survey to as low as 20 samples in a worst case scenario. Once this decision is made, the organization can consider what other information might usefully be gathered during sample collection provided samplers are competent and can recognize the type of overburden sampled. Little or no increase in survey costs and only normal powers of observation are needed to obtain the additional information, for example, soil color.

THE SOIL SURVEY AS PART OF THE EXPLORATION PROGRAM

Soil is normally regarded as all unconsolidated material above bedrock, although it is more correctly that portion of the unconsolidated surficial deposit that is being altered by the physical and chemical processes of soil formation. Thickness of soil varies from mm in some arctic or alpine environments to tens or hundreds of meters in some tropical environments. Soil thickness in glaciated terrains typically averages 1 to 2 metres. Soil development involves climatic factors and organisms as conditioned by relief and water regime. These act through time on geological material to modify its properties and change major and trace element distributions. The resulting soil profile is divided into three layers—the A, B and C horizons. The C horizon constitutes unaltered soil parent material overlying bedrock.

Trace metal content of a soil sample is normally considered to represent a rather limited area. Large numbers of samples therefore need to be collected systematically to evaluate a mineral property. Soil sampling as a reconnaissance technique can be considered where mechanical movement of overburden, as in talus or colluvial deposits, or chemical dispersion in active groundwater flow regimes (i.e., base of slope environments) permit a sample to reflect more than its local environment.

Detailed sampling plans usually follow a square or rectangular grid. Additional samples are taken from landscape environments associated with trace element accumulation, such as depressions or seepage zones, to test for hydromorphic dispersion from a more deeply buried mineral occurrence intersected only by groundwater. In this case, caution must be exercised in the interpretation, as a sufficient number of samples of the same type (i.e., seepage zone samples) must be collected to differentiate between genuinely anomalous and background seepage conditions.

Line spacing and sample interval are controlled by many factors, including:

(1) anticipated size of the mineral occurrence at the bedrock-overburden interface;
(2) local dispersion processes;
(3) geology;
(4) topography;
(5) favorability of the area;
(6) size of the area under investigation; and
(7) availability of funds for personnel and analysis.

Compromises are quite normal; sample density determined on technical grounds alone must be reconciled with availability of time and people. A reduction from optimum parameters usually takes the form of increasing line spacing and/or sample interval.

In Canada, reconnaissance grids commonly comprise lines 300 m apart with a sample interval of 150 m. This can screen an area quickly but significant mineralization may be missed. Ideally, the presence of mineralization is indicated by at least two adjacent samples. However, on a reconnaissance survey, unless the mineral occurrence has a very large surface expression, its presence is only likely to be indicated by single point anomalies. Sample locations are marked in the field in a permanent manner so that followup can commence from a known point. Subsequent detailed soil grids then commonly use a line spacing of 50 to 100 m with a sample interval of 25 to 100 m or perhaps even less if soil anomalies associated with narrow veins are the target.

There is often no obvious relationship between the size and shape of a geochemical anomaly and size and shape of its bedrock source. Consider the Bell Copper deposit of central British Columbia (December, 1980: 53.4 million tons @ 0.51% Cu, 0.34 gm/t Au (Worobec and Needham, 1981)). The ore zone is up to 300 m wide and 100 m long and is surrounded by a 3500 × 2500 m alteration halo (Knauer, 1975). Geochemically the deposit is reflected by a Cu anomaly, in seepage soils downslope from the deposit, at two sites 120 m apart (Figure 3.2). Despite this apparently poor response, the geochemical survey is credited with significantly assisting in the discovery (Carson et al., 1976). The small anomaly and low metal enrichment near this large suboutcropping Cu deposit are a function of a relatively continuous and impenetrable cover of exotic glacial drift 3 to 30 m thick.

AN EXPLORATION EXAMPLE— THE 'QUICK AND DIRTY' VERSUS THE 'SLOW AND PROFESSIONAL' APPROACH

Consider the following case history. A soil survey was conducted in a mountainous region of British Columbia. Results for Mo and Pb (Figures 3.3A and 3.3B) are the only geochemical data available.

> If you were asked to interpret these data and recommend followup procedures, what methods would you use? How would you plan ground traverses to check existing anomalies? Mark your traverse routes on each figure. Figure 3.4 is an airphoto of the survey area provided to assist in your decision(s).

Because of its anomalous Cu and Mo geochemistry, the property was acquired in the late 1960's as a porphyry copper play. Maps presented above comprise a small portion of a 12,000 sample soil grid. The claims were allowed to lapse in 1975 as a result of an adverse political climate, low metal prices, high energy costs (for porphyry Cu deposits) and their remote location. By 1980 political climate and economic conditions had changed dramatically. An all-weather road had been constructed for logging operations and significant Ag and Au occurrences were found to be associated with the Pb anomalies.

> Did your followup recommendations in the preceding paragraph consider the lead anomalies? The scenario described is real and you have as much information as was available to the initial company. The geochemical maps of Figure 3.3 were in the public domain by 1972.

The ability to interpret geochemical information quickly with a view to accurately predicting likely bedrock sources of metals of interest can be a very important factor in acquiring land in a competitive environment.

Examination of the contour plots of Figures 3.3A and 3.3B reveals that apart from highlighting the maximum values, no interpretation is possible. Absence of topographic information requires assumptions be made on the position of geochemical anomalies relative to landscape features on the airphotograph. Each anomalous site has to be revisited prior to interpreting its exploration significance. The original surveyors had not considered this possibility, and revisiting anomalous localities is not possible today—the grid having disappeared. Exploration requires the survey to be rerun in areas of interest. Was this necessary?

Two serious problems are identified:

(1) The position of the soil grid was unknown relative to local topography. This could have been easily avoided using a government issue, topographic map enlarged to an appropriate field scale as a base map. If a government topographic map was not published at the time of

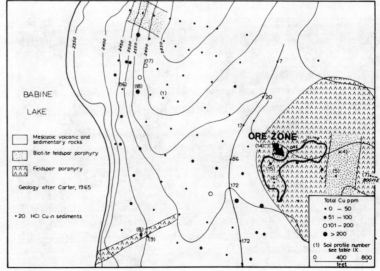

FIGURE 3.2—Bell copper deposit, central British Columbia, Canada, showing distribution of Cu in near surface soils. Note the size and position of the Cu anomaly relative to the ore deposit (modified from Bradshaw, 1975).

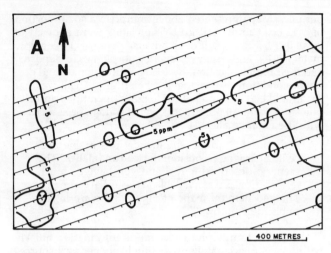

the survey, a topographic map could have been prepared from available airphotographs.
(2) Grid lines were not labelled in the field in a sufficiently permanent fashion.

It is estimated that enlarging the entire area of interest from the government 1:50,000 topographic map to a 1:5,000 field scale would cost $500 today. Preparation of a topographic base map using an orthophotograph could be done for less than $10,000. Cost of permanently labelling grid lines, using 2.5 cm × 5 cm aluminum tags stapled to trees or pickets, is about $0.05 per tag or about $600 total. The extra labor involved in using aluminum tags might increase survey duration and costs by as much as 2% to 5%.

The initial survey was helicopter assisted. At $30 per sample, survey costs amounted to $360,000. Analysis averaged $10 per sample or $120,000. Head office overhead for geochemical interpretation, drafting, project management, etc. is estimated at $200,000. The total budget for the survey is approximately $680,000. Costs that would have been incurred had base maps been prepared in advance and had grid lines been labelled in a permanent fashion can be established using the following figures:

(1) Base maps $500 or $10,000
(2) Aluminum tags $600
(3A) Extra labor @ 2% of field costs $7200
                    or
(3B) Extra labor @ 5% of field costs $18000

Additional costs for the 12,000 sample grid would thus have ranged between $8300 and $28,600, depending on what options were available and necessary at the time. This amounts to 1–4% of the total budget.

With these cost estimates in mind, would you elect to conduct the survey in a "quick and dirty" fashion, or would you elect to undertake the work in a "slow and professional" fashion? Are there any situations you can imagine that would convince you to change your selection?

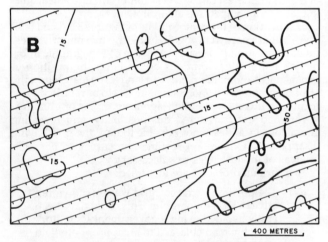

FIGURE 3.3—Public domain geochemical maps for (A) Mo and (B) Pb, central British Columbia, Canada. The anomalies numbered "1" and "2" are referred to in the text.

FIGURE 3.4—Airphoto covering the same area as Figure 3.3.

The geochemical anomalies of Figures 3.3A and 3.3B were followed up over the next several years. Typically an hour or two were needed to locate the first of the anomalous stations. This was due partly to the distribution of landing sites for the helicopter and to poorly marked lines. Commonly the sampler had to walk several hundred meters up and down the line nearest to the landing site to determine grid coordinates on surviving legible flags. This procedure had to be repeated on each line crossed to ensure the correct line was not missed.

Between 25% and 50% of followup effort was spent trying to locate oneself. Still greater inefficiencies accompanied attempts to find anomalies above the treeline where flags tied to rocks had been destroyed by wind.

> If you did not select the "slow and professional" route to conduct your survey in the previous questions, has this illustration changed your mind? The amount of money wasted on the geochemical followup greatly exceeded the maximum $28,600 estimated to undertake the survey properly. Why is it that so many geochemical surveys still repeat many or all of the mistakes described in this example?

The problems were not limited to logistical considerations: speed was considered of the essence and samples were generally of poor quality. For example, the linear Mo feature in the center of the property (No. 1 of Figure 3.3A) occurs, for the most part, along a single line, and it would be reasonable to suspect systematic analytical error. This, however, was not checked before ground followup began. Eventually the line was located and the anomaly found to represent high Mo backgrounds associated with a bog.

> Check the geochemical map to see if you could have determined this from available information. Figure 3.4 is an airphoto of the area and in Figures 3.5A and 3.5B a government topographic map has been approximately superimposed on the geochemical maps of Figures 3.3A and 3.3B. Does this additional information assist your interpretation?

The large Pb anomaly in the east (No. 2 of Figure 3.5B) approximately coincides with talus fans visible in the airphoto. This probably reflects, in part, enhanced background associated with immature soil development compared to forested areas. Location of the source of Pb is straightforward once topographic information at the right scale is in hand.

An extremely important cautionary note is needed with reference to superimposing a topographic map on an existing geochemical map. The procedure is better than nothing for illustrating gross geochemical patterns but is definitely unacceptable as a substitute for preparing a base map in advance and using it to control sampling. Samplers must also be instructed to routinely note, as sampling proceeds, locations of valley bottoms, creeks, mountain tops, cliffs, lakes, swamps and bogs as well as cultural features such as roads, claim posts, houses and dumps. This upgrades the control provided by the map.

Superimposing topographic information, such as the drainage network, on a geochemical map in the absence of ground truth can be a costly mistake. Figure 3.6A represents such an example also derived from the 12,000 sample soil survey. The draftsperson, in isolation from the project manager, transferred information from uncorrected airphotographs enlarged to the same scale as the geochemical map. Creek positions were approximated in the absence of ground truth, a fact not made clear on the maps or to the followup crew.

The Cu–Pb anomaly shown at the head of and following a creek on Figure 3.6A leads to a possible interpretation that differs markedly from what would be suggested if the anomaly had been correctly located on the well drained ridge some 400 meters to the south (Figure 3.6B). Followup, based on Figure 3.6A, focussed on the wrong area on the ground and was fortunate to intersect the edge of the anomaly in a location where metal levels had initially been assumed to be background. A second effort was needed to complete the followup assignment.

This case history illustrates the fundamental importance of topography for interpretation and followup of geochemical anomalies. Government topographic base maps are usually available, and the prudent explorer should, without question, enlarge these maps to the required field scale at a cost of $50 to $100 per map (calculated based on a charge of about $8 per square foot). These should be made available to field crews before sampling begins as their base for controlling the sampling. If published topographic control is poor, an orthophotograph having topographic contours should be prepared despite its initial high cost. A false economy is achieved by assuming these costs can be saved.

GEOCHEMICAL FACTORS AFFECTING TRACE ELEMENT DISTRIBUTION: SOIL DESCRIPTIONS

Soil survey results are susceptible to artifacts (false geochemical patterns, both highs and lows) caused by sample composition. Factors such as soil texture, organic matter, Fe and/or Mn oxides and clay content, proximity to bedrock, bedrock composition, site drainage, pH and soil horizon can, singly or in combination, cause abnormal trace element concentrations unrelated to proximity to mineralization. There is thus no argument in professional circles that proper interpretation of geochemical data requires that not only must samples be collected properly but also detailed descriptive observations be made at the time of sample collection. Furthermore, note taking forces the sampler to look at the material and, once familiarity with proper sampling is acquired, becomes a quick check list to ensure high quality samples are taken.

Project managers often complain that obtaining such data is costly and generally little or no use is made of the information. This is often a valid complaint. Moreover, even when such data are obtained, many exploration personnel inadvertently accept poor quality sampling by others and do not investigate all the parameters contributing to the geochemical patterns being interpreted. Usually, failure to systematically examine the large volumes of data recorded in the field is due to the enormity of the task. However, availability of microcomputers and software for inexpensive, high quality plotting of field parameters should facil-

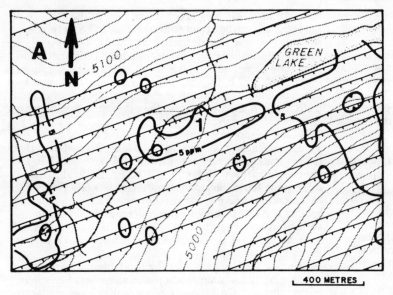

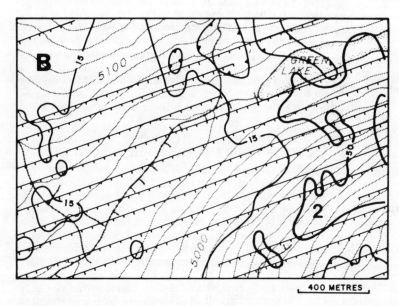

FIGURE 3.5—The geochemical survey of Figure 3.3, but with a government issue topographic map enlarged to the same scale and superimposed. This procedure is acceptable here for demonstration purposes, but is unacceptable for topographic control in an actual working situation.

itate utilization of this information. Remember the objective is to identify false or spurious anomalies before they are followed up in preference to bona fide anomalies.

The following sections describe how both sample composition and landscape environment can be documented in the field and how the resulting information can significantly influence interpretation of geochemical data. Assuming the reader is interested in adopting computer based procedures, an example of a coding form, which has evolved from the author's experience in temperate climatic regions of North America, is provided. The form is based on an 80 column record of field observations in which the first 39 columns contain information of general interest for any type of geochemical survey (Appendix I). The remaining 41 columns describe 23 parameters that can be observed or measured on site (Appendix II). Actual forms are illustrated in Figures 3.7 and 3.8. These allow coding of five (clipboard size) or one (notebook size) sample per sheet, respectively. Forms can be printed on ordinary, multiple copy or waterproof paper.

Selection of codes must be straightforward and not slow sample collection significantly. Codes described here are part of a comprehensive system for all types of geochemical samples and so may, to the casual observer, appear incongruous: this is not the case. Items or categories can be added or deleted as needed. Even those not using a computer will find recording information ensures a high quality of both sample collection and interpretation.

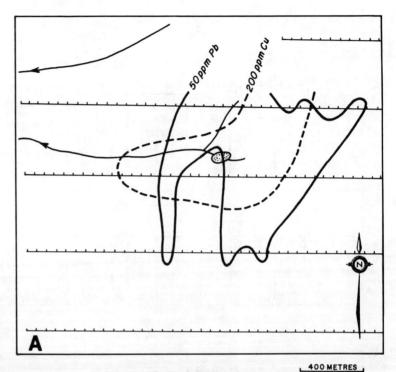

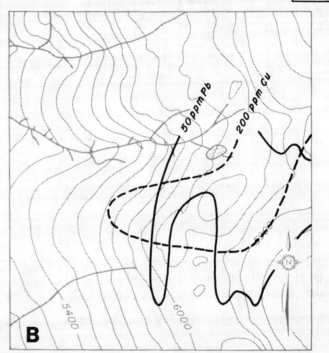

FIGURE 3.6—(A) The practice of freehanding topographic control onto a geochemical map should be avoided. The geochemical anomaly outlined by 50 ppm Pb and 200 ppm Cu contours in this example is incorrectly shown as being at the head of and along the stream—possibly suggesting a seepage anomaly. Followup failed to reproduce the anomaly. (B) Followup did, however, detect an anomaly 400 m to the south associated with a ridge and talus slopes and formed by mechanical dispersion. Example is from central British Columbia, Canada.

Sample Type (Columns 1–2)

Although the form allows for fourteen types of samples (Appendix I), an effort should be made to collect only one type within a survey. Mixing sample types, as in the first example of this chapter, can prove disasterous to the exploration program. Another example illustrates this point further.

A soil survey was undertaken to evaluate a prominent gossan in an alpine environment. Overburden consists of talus debris intermittently stabilized by grass. Soil profile development, involving near-surface leaching and accu-

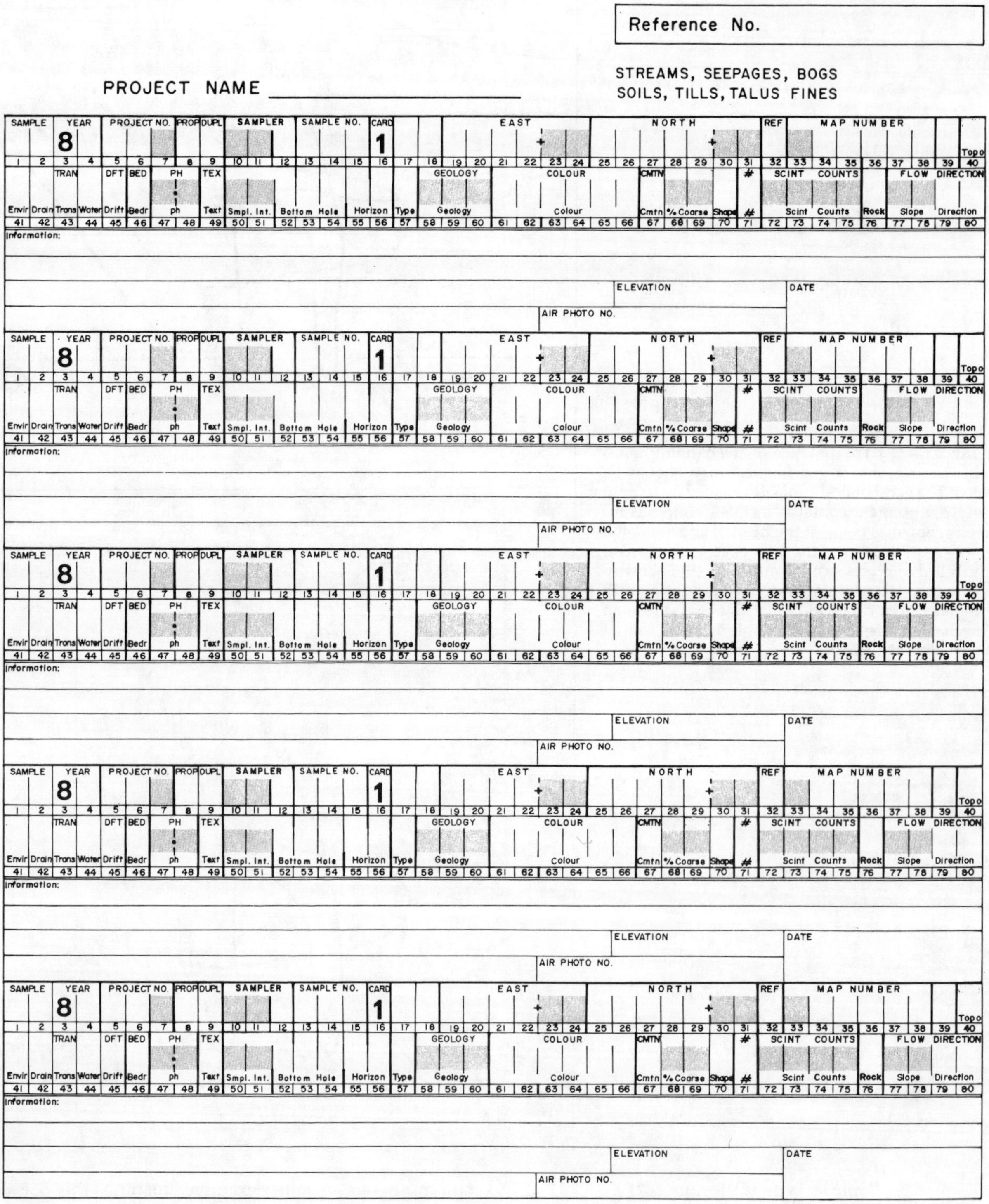
Reference No.

PROJECT NAME ____________________

STREAMS, SEEPAGES, BOGS
SOILS, TILLS, TALUS FINES

SAMPLE | YEAR | PROJECT NO. | PROP | DUPL | SAMPLER | SAMPLE NO. | CARD | EAST | NORTH | REF | MAP NUMBER
8 | 1 | Topo
1 2 3 4 5 6 7 8 9 10 11 12 13 14 15 16 17 18 19 20 21 22 23 24 25 26 27 28 29 30 31 32 33 34 35 36 37 38 39 40
TRAN | DFT | BED | PH | TEX | GEOLOGY | COLOUR | CMTN | # | SCINT COUNTS | FLOW DIRECTION
Envir | Drain | Trans | Water | Drift | Bedr | ph | Text | Smpl. Int. | Bottom Hole | Horizon | Type | Geology | Colour | Cmtn | % Coarse | Shape | # | Scint Counts | Rock | Slope | Direction
41 42 43 44 45 46 47 48 49 50 51 52 53 54 55 56 57 58 59 60 61 62 63 64 65 66 67 68 69 70 71 72 73 74 75 76 77 78 79 80
Information:
ELEVATION | DATE
AIR PHOTO NO.

SAMPLE | YEAR | PROJECT NO. | PROP | DUPL | SAMPLER | SAMPLE NO. | CARD | EAST | NORTH | REF | MAP NUMBER
8 | 1 | Topo
1 2 3 4 5 6 7 8 9 10 11 12 13 14 15 16 17 18 19 20 21 22 23 24 25 26 27 28 29 30 31 32 33 34 35 36 37 38 39 40
TRAN | DFT | BED | PH | TEX | GEOLOGY | COLOUR | CMTN | # | SCINT COUNTS | FLOW DIRECTION
Envir | Drain | Trans | Water | Drift | Bedr | ph | Text | Smpl. Int. | Bottom Hole | Horizon | Type | Geology | Colour | Cmtn | % Coarse | Shape | # | Scint Counts | Rock | Slope | Direction
41 42 43 44 45 46 47 48 49 50 51 52 53 54 55 56 57 58 59 60 61 62 63 64 65 66 67 68 69 70 71 72 73 74 75 76 77 78 79 80
Information:
ELEVATION | DATE
AIR PHOTO NO.

SAMPLE | YEAR | PROJECT NO. | PROP | DUPL | SAMPLER | SAMPLE NO. | CARD | EAST | NORTH | REF | MAP NUMBER
8 | 1 | Topo
1 2 3 4 5 6 7 8 9 10 11 12 13 14 15 16 17 18 19 20 21 22 23 24 25 26 27 28 29 30 31 32 33 34 35 36 37 38 39 40
TRAN | DFT | BED | PH | TEX | GEOLOGY | COLOUR | CMTN | # | SCINT COUNTS | FLOW DIRECTION
Envir | Drain | Trans | Water | Drift | Bedr | ph | Text | Smpl. Int. | Bottom Hole | Horizon | Type | Geology | Colour | Cmtn | % Coarse | Shape | # | Scint Counts | Rock | Slope | Direction
41 42 43 44 45 46 47 48 49 50 51 52 53 54 55 56 57 58 59 60 61 62 63 64 65 66 67 68 69 70 71 72 73 74 75 76 77 78 79 80
Information:
ELEVATION | DATE
AIR PHOTO NO.

SAMPLE | YEAR | PROJECT NO. | PROP | DUPL | SAMPLER | SAMPLE NO. | CARD | EAST | NORTH | REF | MAP NUMBER
8 | 1 | Topo
1 2 3 4 5 6 7 8 9 10 11 12 13 14 15 16 17 18 19 20 21 22 23 24 25 26 27 28 29 30 31 32 33 34 35 36 37 38 39 40
TRAN | DFT | BED | PH | TEX | GEOLOGY | COLOUR | CMTN | # | SCINT COUNTS | FLOW DIRECTION
Envir | Drain | Trans | Water | Drift | Bedr | ph | Text | Smpl. Int. | Bottom Hole | Horizon | Type | Geology | Colour | Cmtn | % Coarse | Shape | # | Scint Counts | Rock | Slope | Direction
41 42 43 44 45 46 47 48 49 50 51 52 53 54 55 56 57 58 59 60 61 62 63 64 65 66 67 68 69 70 71 72 73 74 75 76 77 78 79 80
Information:
ELEVATION | DATE
AIR PHOTO NO.

SAMPLE | YEAR | PROJECT NO. | PROP | DUPL | SAMPLER | SAMPLE NO. | CARD | EAST | NORTH | REF | MAP NUMBER
8 | 1 | Topo
1 2 3 4 5 6 7 8 9 10 11 12 13 14 15 16 17 18 19 20 21 22 23 24 25 26 27 28 29 30 31 32 33 34 35 36 37 38 39 40
TRAN | DFT | BED | PH | TEX | GEOLOGY | COLOUR | CMTN | # | SCINT COUNTS | FLOW DIRECTION
Envir | Drain | Trans | Water | Drift | Bedr | ph | Text | Smpl. Int. | Bottom Hole | Horizon | Type | Geology | Colour | Cmtn | % Coarse | Shape | # | Scint Counts | Rock | Slope | Direction
41 42 43 44 45 46 47 48 49 50 51 52 53 54 55 56 57 58 59 60 61 62 63 64 65 66 67 68 69 70 71 72 73 74 75 76 77 78 79 80
Information:
ELEVATION | DATE
AIR PHOTO NO.

FIGURE 3.7—Geochemical coding form for use with a clipboard to record five samples per sheet. Note that the type of information to be recorded is indicated above or below each box.

TABLE 3.1—Comparison of mean and threshold levels for Cu, Mo and Zn in soils and associated talus fines on andesitic bedrock, central British Columbia, Canada. Aqua regia extraction of minus 80 mesh fraction.

Element		Soils	Talus fines
Mo (ppm)	Threshold	27	22
	Mean	6	5
Cu (ppm)	Threshold	598	963
	Mean	133	225
Zn (ppm)	Threshold	522	388
	Mean	119	112
Number of samples		23	197

higher in the talus fines. This may be related to a Cu occurrence being exposed only in an area of talus fines. Alternatively, and more probably, it reflects more extensive leaching of Cu from B horizon soils relative to unweathered talus fines. Soil and talus fines data have to be reviewed independently and the prospectivity of the property judged after a synthesis of the anomalies defined by both sample types.

Sample Number (Columns 10–15)

The reader will note that six digits (codes 10–15) have been reserved for the sample number (Appendix I) and an all numeric format has been recommended. This ensures data can be handled easily by any computer system. In any large organization there are likely to be a variety of strongly held opinions on the best numbering scheme: to ensure uniformity the company geochemist should have the authority to impose a single system.

The sample number should summarize information in as few digits as possible. Effective long term management of data can be achieved if each sample is coded to include the following information:

(1) Type of sample;
(2) Year of sample collection;
(3) Project code and/or property code;
(4) Sampler identification; and
(5) Sample number.

Two strings of digits are recommended, the first representing an archive code comprising (1), (2) and (3) above (columns 1 through 7 of the form), the second representing the sample number comprising (4) and (5) above (columns 10 through 15 of the form). The maximum number of digits, 7 for the archive number and 6 for the sample number, must remain constant. Sample numbers should not be suffixed with A, B, C, etc. If it is believed that recording too many numbers on a bag will slow sampling, sample bags can be prenumbered with the archive number using a stamp. Failure to follow sample numbering rules leads to increased costs, lost data and slow turnaround. Problems can sometimes be fixed, but this requires a needless, labor intensive solution.

An alternative method of numbering samples involves use of the grid coordinates. This method is preferred by

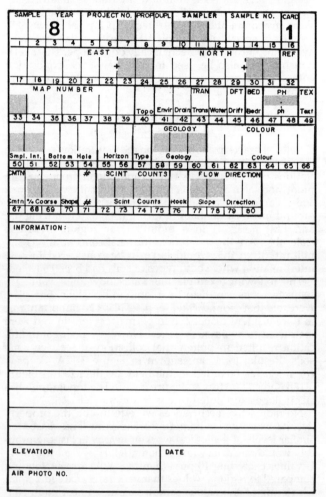

FIGURE 3.8—Geochemical coding form as in Figure 3.7, but designed for use in a geological notebook to record one sample per sheet. This is most appropriate for the geologist who collects soil samples on an occasional basis.

mulation of iron oxides in the B horizon, occurs only on stabilized talus where it is promoted by good drainage resulting from the steep topography and coarse texture of the talus deposits. Sampling comprised collection of fresh talus fines from the upper 5 cm of recent deposits or stabilized talus fines from an iron-rich B horizon at depths of 20 cm to 60 cm beneath vegetation.

Table 3.1 summarizes mean and threshold levels of Mo, Cu and Zn for the two overburden types. For some elements, such as Zn and Mo, distribution of metal appears independent of sample type, whereas for Cu values are

prospectors and some companies. Use of grid coordinate labels can pose serious problems for computer processing and can introduce difficulties in retrieving archived sample pulps.

Whatever scheme is used, sample numbers should not be subsequently changed or modified (for example by field, office or laboratory staff or to make them more compatible with computer processing) as this can lead to irreversible errors.

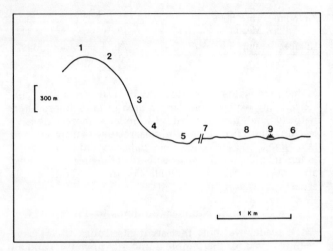

FIGURE 3.9—Schematic summary of the topographic environments described using column 40 of the coding form.

Topography (Column 40)

In the absence of glacial till, the bedrock source of an anomaly located near the top of a mountain (Code 1, Appendix II) is probably immediately beneath or upslope of the anomaly. In contrast, anomalies in base of slope environments (Code 4) may reflect hydromorphic metal accumulation in seepage zones. Sampling of soils in depressions (Code 6) is subject to the same type of metal enrichment as base of slope environments. Figure 3.9 summarizes topographic conditions represented by the computer codes.

Distributions of Ca and V are illustrated in Figure 3.10A and 3.10B, respectively. Soils are developed on residual material or perhaps slightly transported downslope.

> What interpretation can be offered for the source of each element?

Figures 3.10C and 3.10D represent the same information depicted on a government issue topographic map. Calcium enrichment in the range 0.16–0.27% (aqua regia leachable metal) appears to be topographically controlled within a 140 m interval at lower elevations, and highest values (0.5–0.6%) are found in base of slope areas where groundwater might be expected to emerge. Values less than 0.09% characterize the top of the hill. This might reflect a low background content (to an aqua regia leach) in underlying bedrock or extensive leaching of a Ca-rich source—the leached Ca then accumulating in the seepage environment downslope.

If the latter explanation is correct, the source of the Ca within anomalous zones in soils would probably be bedrock 300–500 m upslope. In contrast, highest values of V lie along the axis of the ridge (Figure 3.10D). The source of the V would be interpreted as being immediately below the anomaly or very close by.

Figure 3.11 shows a standard method of following up stream sediment anomalies using bank soils. The base of slope environment beside the creek, upslope of alluvial valley fill, is specifically sampled to locate metal dispersed hydromorphically and mechanically downslope. These relationships can also be used in reconnaissance exploration with soil traverses positioned to take advantage of topography.

Site Drainage and Groundwater Seepage (Columns 42 & 44)

Soils characterized by prolonged saturation are often associated with reducing conditions. Behavior and concentrations of metals in such areas may differ from those in more normal oxidizing conditions found in well drained soils. For example, Table 3.2 summarizes average background metal contents of reducing (BG) and oxidizing (BF and BM) horizons in a mountainous region of British Columbia. Are the differences in metal concentrations significant? If so this would have to be considered in data interpretation where both types of soils had been collected.

The following example illustrates how high values in seepage zones can distort exploration decisions. Samplers were told that groundwater seepage plays an important role in trace element dispersion in permafrost terrain and were instructed to avoid such zones. However, poor communication resulted in preferential collection of seepage zone soils. Results for U are shown in Figure 3.12A. Seepage conditions as recorded in the field are depicted in Figure 3.12B and soil texture in Figure 3.12C. The topography has steep slopes and outcrops are abundant.

No radioactive occurrences were found on the property despite extensive radiometric prospecting and, based on similar levels of geological work in nearby areas, the property would rate a very low U potential. However, the many U values exceeding 10 ppm (Figure 3.12A) are outstanding compared to regional values that very rarely exceed 2 ppm. Clustering of anomalous samples along a linear, fault controlled valley stimulated interest following the geological model described in Chapter 2. Although geochemical notes were available, these were unfortunately not checked until several days of followup had established that overburden was thin and bedrock was exposed intermittently along the valley floor. No radioactive occurrences were found. Had normal soils been sampled initially there would have been no U anomaly and the expense of helicopter assisted followup could have been avoided.

Overburden Origin (Column 45)

Predicting the location of bedrock source(s) of metals depends on recognizing and understanding the genetic significance of the overburden from which the soils are developed and relating geochemical anomalies to geochemical models (Chapter 6). The next few examples illustrate the difficulties encountered if the overburden type remains

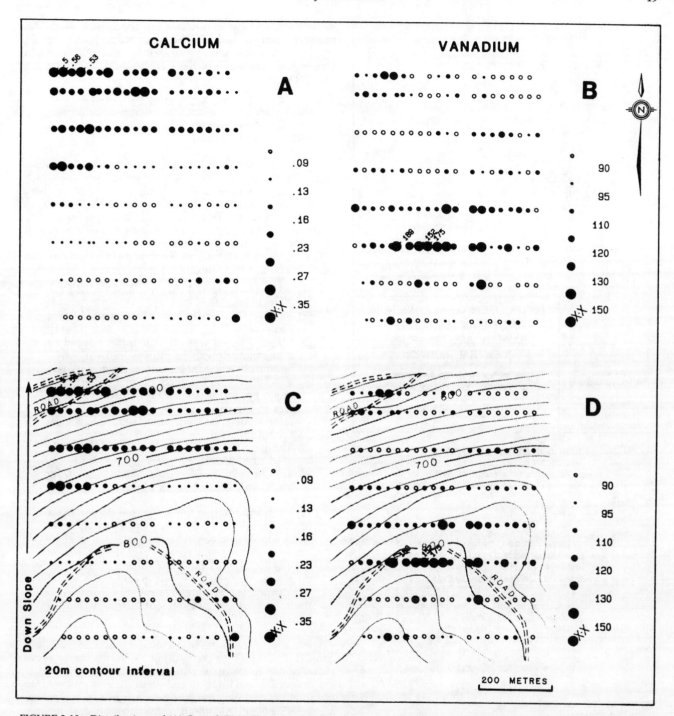

FIGURE 3.10—Distributions of (A) Ca and (B) V. The same data are shown in (C) and (D), respectively, but are plotted on a topographic base map. Note the effect that availability of topographic information has on geochemical interpretation. Property is on Vancouver Island, British Columbia, Canada.

unknown throughout the exploration process. Remember, "difficulty" means extra dollars are spent, and features that should be investigated may remain unrecognized.

A followup survey was initiated to evaluate a northwestward trending, linear Zn anomaly reported in public domain assessment files for a massive sulfide camp. Overburden comprises thin, locally derived till. Previous workers established that the Zn anomaly parallelled a small creek. Followup to the assessment work involved trenching perpendicular to the anomaly at three locations: bedrock was

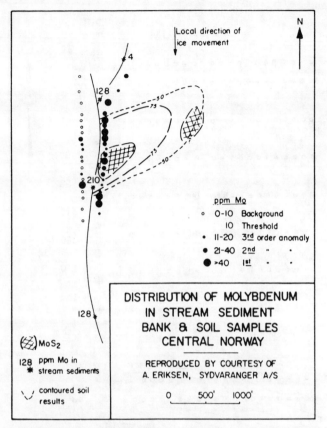

FIGURE 3.11—Followup of a stream sediment anomaly using bank soils. This is a standard practice to locate inputs of anomalous metals to stream sediments as a result of hydromorphic or mechanical dispersion from sources upslope (from Bradshaw et al., 1979).

TABLE 3.2—Average background values of trace elements in well drained and saturated soils overlying granodiorites in mountainous terrain, central British Columbia. Minus 80 mesh fraction, nitric/perchloric acid digestion. Unless otherwise indicated all values in ppm.

Element	Well drained	Poorly drained
Cu	68	80
Zn	34	25
Fe	1.7%	1.4%
Mn	140	170
Pb	4	5
Mo	4	8
Number of samples	170	11

not reached. Several drill holes also tested the linear feature without success.

Workers familiar with the camp and "experienced" in soil sampling were used in a reevaluation of the ground for a new group in 1982. Their report described use of a mattock or grub hoe to sample to a depth of 10 to 15 cm (4 to 6 inches) and retrieve a "constant red-brown soil which did not vary from site to site". The geochemical pattern for Zn repeated the assessment report work and is illustrated in Figure 3.13. Locations of the trenches and drill holes are also shown.

> Would you proceed any further with evaluation of the Zn anomaly? What interpretation would you give the Zn enhancement, and are the high Zn values really anomalous?

In this case tenure on the ground was allowed to lapse. However, the report of a "constant red-brown soil which did not vary from site to site" should provoke skepticism. In most environments this is highly unlikely since some variation is always to be expected. Moreover, subsequently learning that sampling had been accomplished at a rate of 125 samples per man day should fuel suspicions.

The following observations were made in a study undertaken to confirm their results:

(1) Soil pits averaged 4 cm wide and varied from 10 to 15 cm deep.
(2) Pits were generally immediately beside grid pickets.
(3) Where a picket was on a mound produced by a tree blow down, the pit was in the mound.
(4) Several old roads are present on the grid. Where a picket was on the road, or on an overgrown mound produced by road construction, the pit was in the road or in the mound. To be fair, reforestation had somewhat disguised the location of the road.
(5) The anomaly coincided with a 50 to 100 m wide alluvium-filled valley.
(6) Trenches mainly intersected alluvium, till being found only at the bottom of the trenches.

> With this information in hand, what do you think of the quality of the 1982 survey? Have the relatively simple field observations suggested an alternative geochemical interpretation of the data? From the results, predict the location of a suboutcropping massive sulfide deposit grading 12% Zn. If you cannot "see" the massive sulfide occurrence in the survey data, would you recommend the survey be repeated?

This example continues in the section describing soil horizons. It illustrates how recognition of overburden type can radically alter followup recommendations. The initial workers believed overburden to comprise till and followed up the anomaly assuming the northwesterly linear represented a mineralized horizon meriting trench and drill testing. Their failure resulted because alluvium was not initially distinguished from till.

Consider another example. A soil survey was initiated to evaluate a favorable geological environment associated with a large number of weak anomalies from an airborne electromagnetic survey. However, geophysical responses where the survey overflew a large lake gave similar responses that were interpreted as being due to clay-rich lake bottom sediments, of no interest to the exploration program.

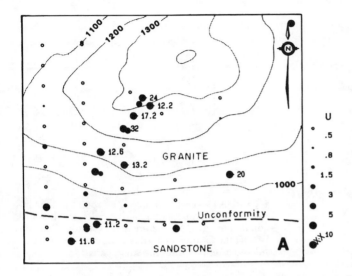
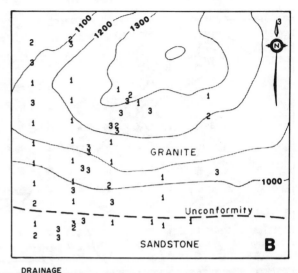

DRAINAGE
1 — DRY 2 — MOIST 3 — SATURATED

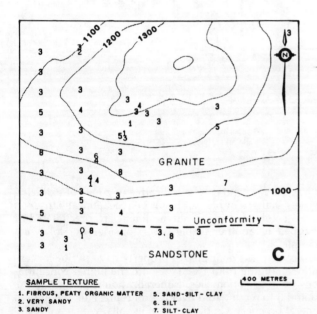

SAMPLE TEXTURE
1. FIBROUS, PEATY ORGANIC MATTER
2. VERY SANDY
3. SANDY
4. SAND-SILT
5. SAND-SILT-CLAY
6. SILT
7. SILT-CLAY
8. CLAY

400 METRES

FIGURE 3.12—(A) Distribution of U (ppm), (B) field observations of seepage conditions, and (C) field estimates of sample texture. Note the strong correlation between high U values, seepage areas and soils having fine textures.

Did the geophysical anomalies on dry land reflect a similar source or possible bedrock targets?

Ground inspection indicated overburden had a fine texture but failed to recognize its origin. However, the soil survey outlined homogeneous enhancement of Pb and Ni in a zone surrounding the lake (Figure 3.14A and 3.14B). These high values are interpreted to reflect lacustrine overburden. The soil survey cannot be used directly to evaluate underlying bedrock, but it does indicate an area where soil geochemistry is not an effective exploration method. The survey also assists in explaining the EM anomalies by attributing them to conductive lake sediments. Other techniques would be needed to evaluate the area.

In the next example the locations of Cu and Mo anomalies and the direction of glacial transport are shown (Figure 3.15). The shapes of the anomalies are consistent with dispersion in the reported glacial direction.

If this was your exploration program, would you refer to the field notes to determine if the reported geochemical interpretation is correct?

Anomaly followup noted that outcrops were abundant and overburden was 0.5–2 meters thick. Boulders in the

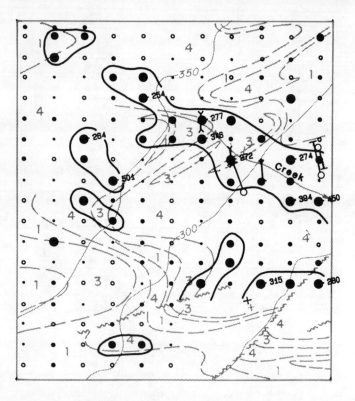

FIGURE 3.13—Distribution of Zn (ppm) from the initial survey in an area having volcanogenic massive sulfide potential, New Brunswick, Canada. Can you predict the location of a suboutcropping massive sulfide occurrence within the survey area?

anomalous soils were found to be angular and were of the same rock type at any one location. Reexamining Figure 3.15, it can be seen that the majority of anomalous conditions are at the highest elevations on the property.

Can an alternate hypothesis be suggested based on the ground truth information?

Another possibility is suggested by the field observations. Overburden within anomalous areas is described as residual or "residual-like" having been transported no more than 10 or 20 m. The source of metal therefore underlies or lies immediately upslope or up ice of the metal-rich zone. The metal-rich zone might not even represent an anomaly, enhanced metal levels simply reflecting different overburden types with higher backgrounds associated with residual soils compared to lower values in glacial tills at lower elevations. Again access to easily collected field observations has significantly affected the geochemical interpretation.

Soil pH (Column 47–48)

The procedure described in Appendix II for measuring soil pH is suitable for use in a field laboratory. It assumes that only approximate pH values are needed to interpret its effect on trace element distributions. Soil pH in temperate climates typically averages 5.5 to 7.5. Values of 4.0 or less are rare and can be indicative of zones of oxidizing pyrite. Changes of pH can result in geochemical barriers if metals solubilized in one environment are precipitated or adsorbed in another (Table 3.3). Consider the following example.

Volcanic rocks on a Cu–Au property have pyrite concentrations ranging from 1% to 5% with the higher concentrations giving IP anomalies. Mineralization grading 1 to 2% Cu and 3 to 6 gm/t gold over 3–4 m represents the known mineral prospect. Cu distribution in soils, which are residual, is shown in Figure 3.16A.

What followup procedure, if any, would you recommend based on the distribution of Cu anomalies?

Soil pH was determined on all samples and a northwesterly trending zone, some 500 m wide, was outlined with values of 3.8—4.2 (shaded areas in Figure 3.16B). This is unusually acidic and probably reflects a zone of oxidizing pyrite in underlying bedrock. Acidic soils lie upslope of several of the Cu anomalies, which are found in areas of near neutral soils, suggesting the possibility that Cu is accumulating at a geochemical barrier.

What procedures would you use in evaluating Cu anomalies on this property?

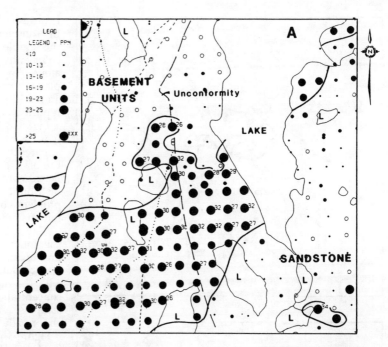

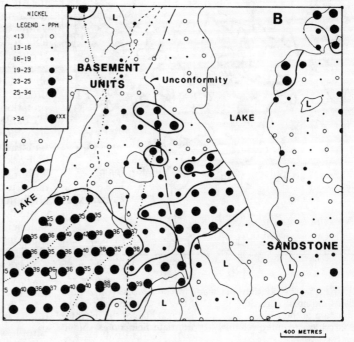

FIGURE 3.14—High values of (A) Pb and (B) Ni indicate the distribution of overburden derived from ancient lake bottom sediments. In INPUT EM surveys this material shows up as a conductor giving a response similar to that found over the present day lakes. Because mixing of overburden (pedoturbation) by frost boil action has destroyed primary structures, the geochemical patterns are a convenient method of delineating these ancient lake bed sediments and discounting the associated EM conductors. The example is from an area of permafrost terrain in the central Northwest Territories, Canada.

Geochemical features developed as a result of pH conditions are relatively inexpensively identified. Table 3.3 indicates that many elements are susceptible to leaching and can accumulate at pH barriers. Potential immobilization factors that often accompany changes in pH include changing Eh conditions and changes in the abundance of substrates, such as Fe/Mn oxides and organic matter, which scavenge trace elements.

Examples of trace metal accumulation with Fe/Mn oxides precipitated as a result of changing pH-Eh conditions in soils are compared to background values for the same area in Table 3.4. In some samples only one element is enhanced (e.g., Co in Sample 1), whereas in other cases several elements are present in abnormally high concentrations (e.g., Mo, Pb, Zn, As and Co in Sample 17). Although statistical methods, such as regression analysis, might be used to

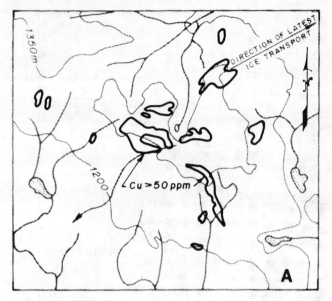
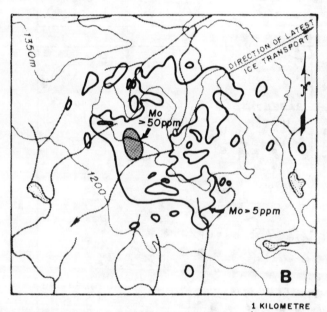

FIGURE 3.15—Distributions of Cu and Mo on a property in central British Columbia. Direction of glacial transport is indicated. Determination of the genesis, glacial versus residual, of the overburden is critical to interpretation. Appropriate field notes should make this possible.

correct for abnormal accumulations, it is obviously better to avoid interpretational problems by not collecting material of unusual composition. In this case, Fe/Mn-rich samples could have been recognized by their strong black or rusty colors.

Texture (Column 49)

Sample texture is determined as the sample is being placed in the sample bag. Identification of very fine or very coarse textures suggests problems are likely in interpretation, and the sampler is advised to try sampling at a different site. If the sample is very fine textured (i.e., clay rich), it may have sufficiently high background metal values to appear anomalous compared to other soils—a "false" anomaly. If the sample is coarse textured and consists largely of clean quartz and feldspar, relatively low metal abundances may be reported despite proximity to a mineral occurrence.

Figure 3.17A shows soil textures for a U property in the Northwest Territories. Fine textured and organic-rich soils apparently control distribution of many high U values (Figure 3.17B). If these data are ignored (Figure 3.17C) insufficient samples remain to adequately define geochemical distributions for followup by radiometric prospecting. The problem could have been avoided during sampling with minimal effort.

Clay-rich samples can sometimes be suspected from analysis following acid digestion. Metal levels associated with fine textured materials frequently exceed those of coarse textured soils as a result of the greater surface area reacting with the digesting acids. Aqua regia leachable aluminum concentrations, for example, should be elevated in clay-rich samples, and spot highs could be interpreted as reflecting abnormally clay-rich material.

Figure 3.18 represents a portion of a soil survey exploring for Au and base metals. Soluble Al content in Figure 3.18A has a number of high spot values (i.e., single, very high values adjacent to much lower backgrounds). These probably reflect samples having a high clay content and consequently scavenging capability. Figure 3.18B shows coincident Cu features probably representing false anomalies. Creation of these sample-related anomalies could probably have been avoided during sampling and interpretation correspondingly simplified.

Coarse textured material giving a depressed geochemical signal is seen in the soil profile of Figure 3.19. Copper contents in the A and B horizon behave predictably, but as bedrock is approached two sandy layers are present, one within 15 cm of mineralized bedrock (0.15% Cu). Both contain Cu contents of less than 50 ppm. Low Cu values are attributed to the high content of quartz and feldspar sand content, which is unable to retain Cu passing through the sand lenses in solution. The sampler is thus advised to shift the sample location to retrieve a better sample.

Sample Depth (Columns 50–54)

Only an approximate sample depth is needed and an estimate, relative to the length of a shovel blade or some other implement, is adequate. How important to exploration is the sample depth estimate? Consider this example. Samplers trained in arctic Canada were sent to the south coast of Newfoundland to conduct a soil survey. Soils in the arctic are generally thin and immature whereas profiles in southern Newfoundland are likely to be deeper. The samplers recorded that they had collected material from the top of the B horizon at depths from 25 cm to 30 cm. This reflects proper sampling in the arctic, but was it appropriate for Newfoundland?

An example of traverse results from part of a line of samples taken at 50 m intervals is shown in Table 3.5. Orientation studies showed the soil profile had the following characteristics:

TABLE 3.3—Mobility of some of the common trace elements in the surficial environment (based on Levinson, 1980).

Element	pH conditions			Immobilization factors			Heavy minerals
	Acid <5.5	Neutral pH 5.5–7.0	Alkaline pH >7.0	Fe/Mn oxides	Organic matter	Other	
Antimony	Low	Low	Low	Yes		Sulfide; reducing conditions	
Arsenic	Medium	Medium	Medium	Yes		Sulfide; clay conditions	
Barium	Low	Low	Low			Sulfate; reducing conditions; carbonate; clay	
Beryllium	Low	Low	Low	Yes	Yes	Clay	
Bismuth	Low	Low	Low	Very		Reducing conditions	
Boron	V. High	V. High	V. High				Tourmaline
Cadmium	Medium	Medium	Medium			Reducing conditions	
Cerium	Insoluble	Insoluble	Insoluble				Rare earth minerals
Chromium	V. Low	V. Low	V. Low				Chromite
Cobalt	High	Medium to Low	V. Low			Sulfide; adsorption	
Copper	High	Medium to Low	V. Low	Yes	Yes	Sulfide; adsorption	
Fluorine	High	High	High			CaF_2, adsorption	
Gold	Immobile	Immobile	Immobile				Native gold
Iron-Fe^{+++}	High to V. Low	V. Low	V. Low	Yes			Magnetite
Iron-Fe^{++}	High	Medium to Low	V. Low	Yes		Oxidizing condition	
Lead	Low	Low	Low			Insoluble carbonate, sulfate, phosphate; reducing conditions	
Lithium	Low	Low	Low	Yes		Clays	
Manganese	High	High	High to V. Low	Yes		Clays	
Mercury (aq)	Medium	Low	Low	Yes		Sulfide	Cinnabar
Mercury (vap)	High	High	High				
Molybdenum	Low	Medium	High	Yes		Adsorption; in presence of Pb, Fe, Ca, carbonate; sulfide; reducing conditions	
Nickel	High	Medium to Low	V. Low			Sulfide, adsorption, silicate minerals	
Niobium/ Tantalum	Insoluble	Insoluble	Insoluble				Yes
Platinum	Insoluble	Insoluble	Insoluble				Native platinum
Radium	High	High	High	Yes	Yes	Coprecipitation with Ba, Ca, Fe, Mn	
Radon	High	High	High			Limited by half life	
Selenium	High	High	V. High	Yes		Reducing conditions, adsorption	
Silver	High	Medium to Low	V. Low	Yes	Yes	Reducing conditions; sulfide; precipitated by Pb, Cl, chromate, arsenate	
Tellurium	V. Low	V. Low	V. Low				Yes
Thorium	V. Low	V. Low	V. Low			Adsorption by clay, aluminum hydroxides	Yes
Tin	Insoluble	Insoluble	Insoluble				Cassiterite
Tungsten	Insoluble	Insoluble	Insoluble	Yes			Wolframite Sheelite
Uranium	Low to Medium	High	V. High	Yes	Yes	Reducing conditions; special ion precipitates; adsorption	
Vanadium	High	High	V. High			Silicate minerals; reducing conditions, adsorption	V-magnetites
Zinc	High	High to Medium	Low to V. Low	Yes	Yes	Sulfide, precipitated by high carbonate, phosphate	

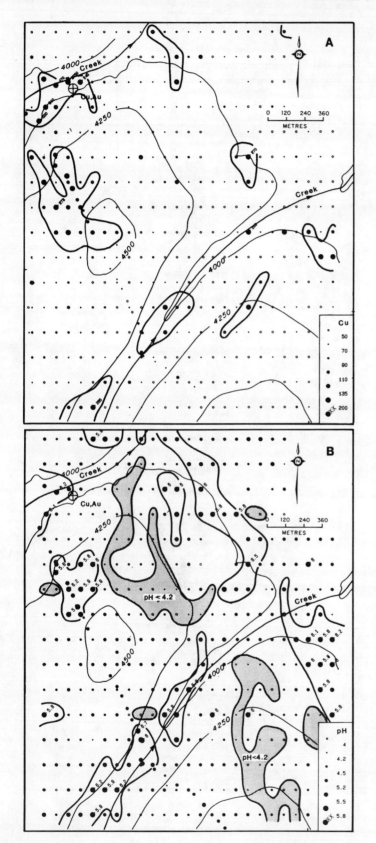

FIGURE 3.16—Distributions of (A) Cu (ppm) and (B) pH values in soils on a property in central British Columbia. Shaded areas in (B) have pH values < 4.2 as a result of oxidation of pyrite in bedrock. Cu anomalies (Cu > 90 ppm) are found downslope of these zones where less acidic sites form geochemical barriers and promote accumulation of Cu.

TABLE 3.4—Accumulation of base metals and pathfinder elements in response to high concentrations of Fe (>10%) and Mn (>10,000 ppm) in soil samples selected from a 1034 sample study, Newfoundland, Canada. Average background values based on all 1034 samples are provided for comparison.

Sample number	Mo ppm	Cu ppm	Pb ppm	Zn ppm	As ppm	Co ppm	Fe %	Mn ppm
1	3	3	24	34	6	20	2.6	10300
2	6	5	18	80	5	48	5.1	13100
3	7	36	46	64	52	93	7.7	23500
4	10	8	22	40	41	44	10.0	121000
5	13	9	58	61	134	266	10.3	74000
6	2	49	20	174	79	29	10.5	530
7	1	22	15	114	8	30	10.7	840
8	12	60	20	193	85	35	11.5	2350
9	8	6	18	70	75	24	5.2	16200
10	1	23	21	93	37	26	11.1	2420
11	9	15	21	111	57	61	10.3	8900
12	1	50	29	100	13	21	11.7	1070
13	18	10	63	97	24	37	10.7	1430
14	14	5	26	44	358	53	19.2	22400
15	48	7	46	24	316	48	16.1	10800
16	16	6	63	74	218	114	40.9	3200
17	51	7	70	141	224	53	12.6	24200
Average (for 1034 samples):	3	8	15	24	17	6	1.6	690

Top horizon	A black, organic-rich zone 10 cm thick.
Second horizon	A light gray to white mineral horizon 10 cm thick.
Third horizon	A light to medium brown mineral horizon 20 cm thick.
Fourth horizon	A lower horizon with a strong rusty color averaging 5 cm + thick; sometimes overlain by a black layer 1 or 2 cm thick. Boulders prevented further penetration.

Was the sampling program acceptable? Samplers trained in one part of Canada had been asked to sample in another environment for which they had no experience. They collected material from what they believed to be the proper horizon, but Mn and Fe contents near their detection limits indicate that at some sites sampling failed to penetrate a leached horizon. At other sites much higher Mn and Fe contents, accompanied by anomalous levels of U, Pb and other base metals, were found when the samplers obtained material from below the leached zone. Exceptionally high As values and the multielement signature at these sites attracted great interest until it was realized that they were artifacts of sampling. Depth of sampling can thus be a guide to the type of soil (horizon) being collected.

Soil Horizon (Column 55–56)

Horizon classifications developed by pedologists vary with country. In North America two major systems are used, one in Canada (Canada Soil Survey Committee, 1978) and the other in the United States (Soil Survey Staff of the Soil Conservation Service, 1975). Table 3.6 compares abbreviations and relevant properties of the major soil horizons. It also summarizes horizon designations used in the United States prior to 1975 (Soil Survey Staff, 1951) because they are still widely used.

Soil horizons represent stages in the continuous physical and chemical variation of soils in time and space (Figure 3.20). Field identification requires observation on the sequence of layers (horizons) down holes and their texture and color to at least the depth sampled. A clue to the changing character of the soil can also be provided by the vegetation (Figure 3.21).

Figure 3.20 indicates that soil horizon sequences change with time and reflect the pedogenic processes of eluviation (leaching) and accumulation of soil components. Rates of change are influenced by climate, landscape and the nature of the soil parent material, with the rate of soil development being aided by factors that promote leaching. In temperate climates the principal processes involved are:

(1) Accumulation of organic matter at the surface and its decomposition to humic and fulvic acids and other complex organic substances.
(2) Downward leaching of humic and fulvic acids and, depending on mineral stability and order of solubility, the alkalies, alkali earths and iron and manganese.
(3) Downward migration of clay minerals, particularly after removal of soluble alkali and alkali earths.
(4) Accumulation of clays and precipitation of dissolved materials lower in the profile. For example, in dry regions alkaline, carbonate-rich horizons can develop. In contrast, precipitation of Fe and Mn oxides at the top of the B horizon characterizes soils developed under higher rainfall conditions.
(5) Hydromorphic dispersion of constituents remaining in solution. These may reappear and

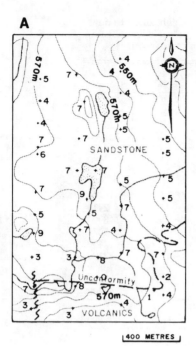

SOIL TEXTURE

0 Organic muck
1 Peat
2 High sand
3 Sand
4 Sand-Silt
5 Sand-Silt-Clay
6 Silt
7 Silt-Clay
8 Clay
9 Gravel

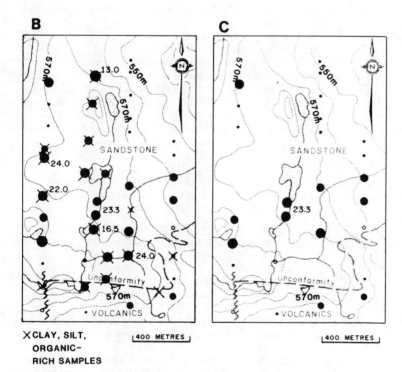

FIGURE 3.17—(A) Field measurement of sample texture; (B) distribution of U (ppm); and (C) distribution of U (ppm) with results from finely textured soils omitted. Too few data remain to evaluate the property and resampling would be required. Property is in the Northwest Territories, Canada.

accumulate elsewhere on the geochemical landscape, for example, in seepage zones.

The importance of recognizing different soil horizons and selecting those which optimise geochemical response to mineralization and minimize sampling variability is not appreciated by many explorationists. Figure 1.4 illustrates the differences in distribution of Mo resulting from poor sampling, using the "sampling at constant depth" philosophy, compared to thorough sampling using a constant soil

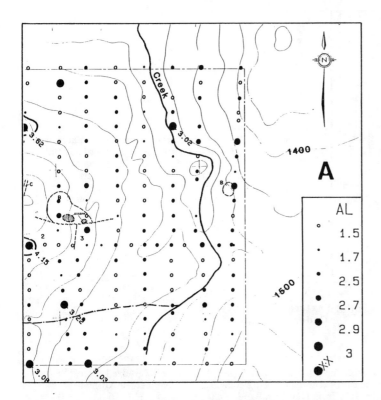

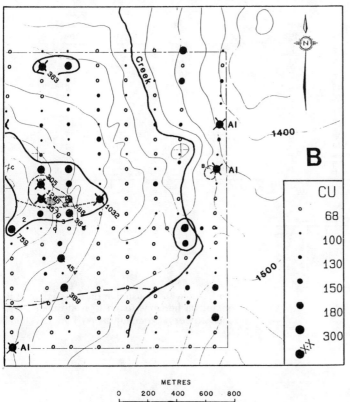

FIGURE 3.18—(A) Distribution of aqua regia soluble Al (%), and (B) distribution of Cu (ppm) with organic-rich samples indicated by an "×" through the sample point. Measles-like patterns for Cu suggest that erratic high values are caused by clay- and organic-rich samples giving high background values.

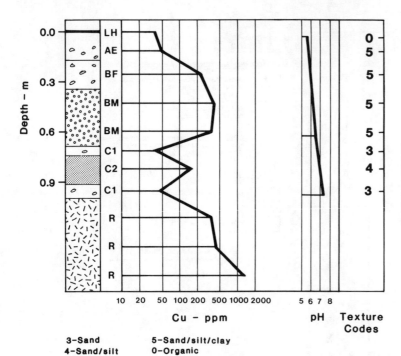

FIGURE 3.19—Soil profile showing the effect of sample texture and soil horizons on Cu content. Note low Cu contents of sandy horizons immediately overlying bedrock containing 0.15% Cu. This reflects inability of sand grains to retain Cu released by weathering. Profile is from south-central British Columbia, Canada.

3-Sand 5-Sand/silt/clay
4-Sand/silt 0-Organic

TABLE 3.5—Results from a single traverse line with soils, sampled at a depth of 25 cm, giving mixed high and low Fe and Mn abundances as a result of AE and BF horizons being collected at different sites. Normal soils contain 300 ppm Mn and 3 to 4% Fe. Example is from Newfoundland, Canada.

Sample/horizon	Mo ppm	Cu ppm	Pb ppm	Zn ppm	Ag ppm	Mn ppm	Fe %	As ppm	U ppm
1 AE	1	1	2	1	0.2	5	0.14	3	5
2 AE	1	1	4	2	0.2	3	0.10	2	5
3 AE	1	3	2	8	0.2	30	0.80	5	5
4 AE	9	18	82	43	0.1	142	1.45	6	5
5 BF	57	14	141	69	0.6	14900	28.9	2390	31
6 BF	3	2	13	17	0.1	1960	2.85	39	5
7 AE	15	24	86	17	1.3	129	0.74	38	41
8 BF	1	6	11	39	0.4	38	2.84	53	5
9 BF	8	34	23	142	0.4	12500	20.30	430	15
10 BF	1	4	6	33	0.1	272	3.39	25	5
11 BF	1	1	7	6	0.1	1115	3.94	42	25

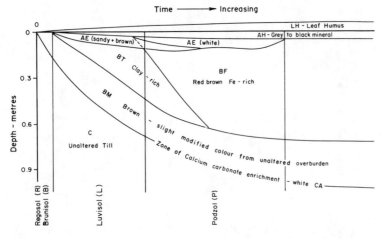

FIGURE 3.20—Relationship in time and space of the common soil horizons found in well drained environments (personal communication, L.M. Lavkulitch, 1971).

TABLE 3.6—Comparison of soil horizon nomenclature simplified from the Canadian (Canadian Soil Survey Committee, 1978) and American (Soil Survey Staff of the Soil Conservation Service, 1975) classifications. Horizon designations used in the American classification prior to 1975 (Soil Survey Staff of the Soil Conservation Service, 1951) are included for reference purposes.

This text	Canadian (1978)	American (1975)	American (1951)	Description
LH	LH	O1,O2	A00, A0	Leaf humus, undecomposed vegetation on the surface above mineral-rich horizons
AH	Ah	A1	A1	Dark brown, gray to black organic-rich mineral horizon. Commonly <15 cm thick
BH	Bh	B2H	B2	Dark brown to black, organic-rich (+ Fe-rich) mineral horizon. Usually at depths >15 cm
AE	Ae	E	A2	Gray to white (occasionally brown) mineral horizon, near surface, usually sandy; accompanied by BF or BT horizon at depth
BF	Bf	Bs	B2	Red-brown, Fe-rich accumulation zone
BT	Bt	Bt	B2	Brown, clay-rich accumulation zone
BM	Bm	B2	B2	Brown horizon only slightly modified in appearance from parent material
BG	Bg	G	Bg	Water saturated zone most of the year; characterized by mottles
CA	Cca	Cca	Cca	White calcium carbonate precipitate in C horizon
C1,C2,..	IC,IIC,..	IC,IIC,..	IC,IIC,..	Parent material
O1,O2,..	O1,O2,..	O1,O2,..	O1,O2,..	Organic-rich bog samples

horizon. It is essential to instruct samplers on proper methods of collection and inform them to collect material from a particular horizon.

Data in Figure 3.22, showing part of a larger grid, were obtained after instructing samplers to take material from the BF horizon. BF and BM horizons were the principal horizons described in the field, but a BH (organic-rich) horizon was also present. This was specifically to be avoided. Fortunately fieldnotes were taken and obvious BH horizons were recorded at three of the approximately 800 sample sites on the complete grid. There are two such sites in the southern half of the survey (Figure 3.22A). However, various shades of dark brown and gray brown to black soils were recorded in addition to the normal browns and red browns (Figure 3.22B). These dark colors, at 43 of the 800 sites, are not consistent with the definition of the BF, BM, BT or BG horizons but are indicative of organic-rich materials. Anomalous accumulation of Cu, probably due to organic matter scavenging, characterizes about one-third of the affected sites (indicated by an × over the large dots on Figure 3.22C). Without the soil color information to provide a check on accuracy of the soil horizon determination, inter-

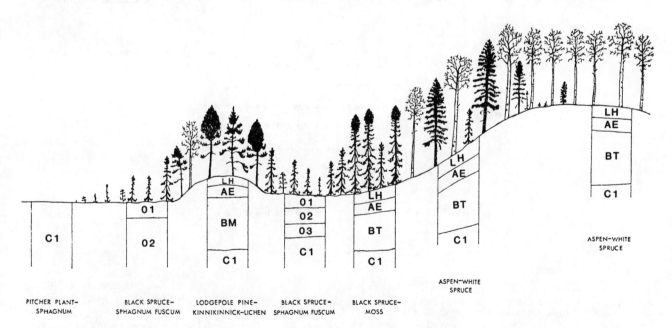

FIGURE 3.21—The underlying soil catena is reflected by the changing character of vegetation supported by different soil types.

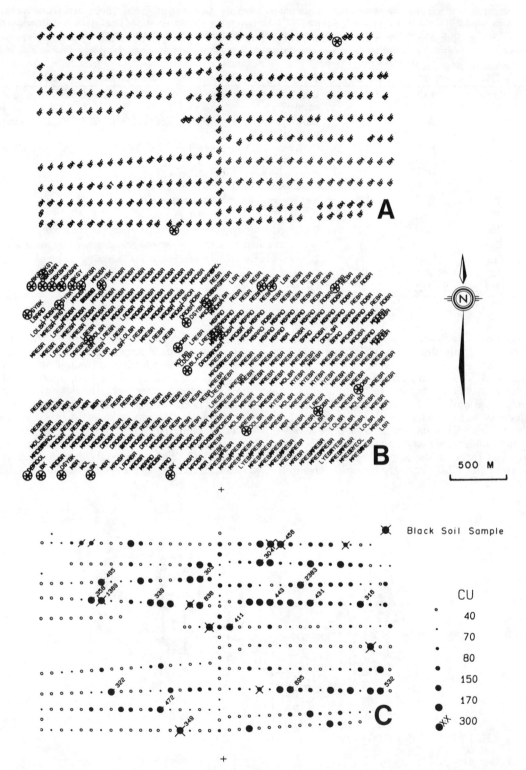

FIGURE 3.22—(A) Soil horizons identified by samplers following the codes described in Appendix II of Chapter 2. (B) Soil color coded with L (light), M (medium) and D (dark) followed by the first two letters of the color. (C) On the basis of (A) and (B), Cu anomalies with an "×" through the symbol are interpreted to be false, resulting from scavenging of Cu by organic matter (from Hoffman, 1985).

pretation and followup recommendations could have been misguided. Subsequent investigation confirmed careless sampling procedures had been used and that proper BF horizon material could have been collected at many of the sites by either digging deeper or shifting the sample site by 5 or 10 m.

Multielement analysis can, on occasion, also assist in predicting problems introduced by sampling. The false anomalies outlined in Figure 3.22 were recognized relatively easily by dark soil colors. Figure 3.23A presents Ca data for the same grid. Aqua regia leachable Ca contents exceeding 0.66% are considered anomalous and occur at 37 of the 43 black soil sites. Strong correlation of high Ca contents with black samples suggests high Ca content could be used to indicate organic-rich soils where soil color alone is not diagnostic. In this way it can be predicted that organic-rich materials were probably collected at an additional 5% of sites. Figure 3.23B illustrates that high Mn contents are also related to dark colored soils at 27 of the 43 sites and that Mn could be used in a similar manner to Ca.

Soil sampling problems identified from Ca and Mn data are reflected by "measles-like" distribution patterns for base metals. In this case about 10% of the samples are adversely affected by poor quality sampling and one third of the Cu anomalies are probably false and not worthy of followup. Remedial action is needed to: (1) prevent recurrence of the sampling problem; and (2) better direct followup to bona fide features.

In the section on overburden origin, an example was cited of a Zn anomaly following an alluvium-filled valley (Figure 3.13). The bedrock source of the Zn would be predicted to lie upstream or to the northwest of the linear soil anomaly. However, because sampling problems were suspected in the initial survey, it was decided to test this interpretation. A series of additional samples were collected, within 1 m of the original sites, on five lines. Both sets of samples were then sent to the laboratory for analysis in the same batch. Figure 3.24 compares results of the two surveys. Significant differences are apparent, both in average metal levels and in the distribution of anomalous results. Subsequently, a

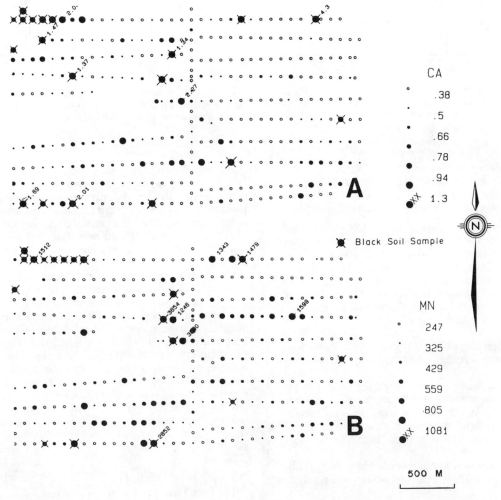

FIGURE 3.23—Distributions of (A) Ca (%) and (B) Mn (ppm) with locations of organic-rich "black" samples shown by an "×" through the sample point (from Hoffman, 1985).

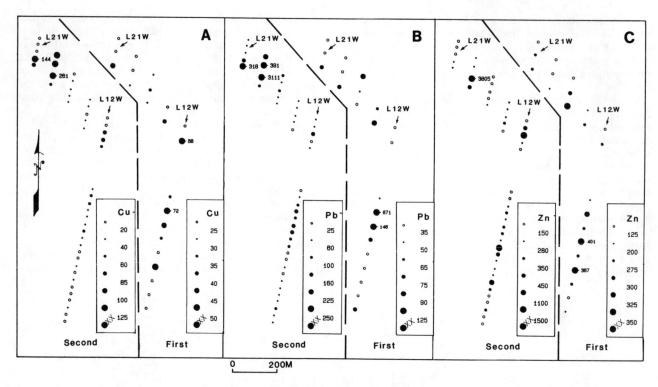

FIGURE 3.24—Comparison of two geochemical surveys for (A) Cu, (B) Pb and (C) Zn. Samples were collected at the same site for both surveys, but the "speed" sampling used in the first survey failed to indicate the suboutcropping volcanogenic massive sulfide deposit found as a result of the anomalies on line L18W of the second survey. All values are in ppm. Modified from Hoffman, 1985.

massive sulfide occurrence was located under 2 m of overburden by followup of the Cu–Pb–Zn anomaly, interpreted to be in residual overburden, along L18W of the second survey. This anomaly is missing from the first survey.

With respect to the data in Figure 3.24, what went wrong with the first survey? Samplers were experienced, having collected tens of thousands of samples in their careers. They were fast, reducing the costs of the exploration company. No fault could be found with the analytical work. On the basis of these results and Figure 3.13, can you list the deficiencies of the first survey?

(1) Leached samples (AE) were taken instead of the iron-rich (BF) horizon. Sampling was too shallow.
(2) Near surface organic matter from the bog overlying the massive sulfide occurrence was sampled. Adjacent samples were disappointingly low in metal due either to collection of AE material or organic material (site specific notes were not taken).
(3) Information was not recorded to predict problem areas where remedial action was needed.

In this case the consequences are dramatic. The initial survey outlined numerous anomalies, but none were associated with the zone of economic significance. Targets defined by the first survey were followed up but did not lead to a discovery. In the second survey, recognition of the complexities of soil development was the basis for sampling the correct horizon, albeit at a greater cost. This led to discovery of the massive sulfide and rejection of the anomaly associated with alluvium. Cost effectiveness was achieved with the proper procedures.

Having confirmed the presence of sampling problems in the initial survey, it was decided to resample the southeast part of the survey. Contour levels chosen on the basis of the original survey of Figure 3.24 were used to outline anomalous conditions in Figures 3.25 A, C, and E for Cu, Pb and Zn, respectively. Figures 3.25 B, D and F were prepared on the basis of the repeat samples. The geochemical legends on the two sets of maps are markedly different, and the anomalous areas are much more extensive using thresholds based on the original, unreliable data. The reader will appreciate that very different interpretations and subsequent exploration strategies would be based on the two sets of maps. Once again, this illustrates the importance of sampling the correct horizon.

Rock Type (Column 58–60)

The importance of rock type as a control on geochemical distributions in soils has already been noted. Trace element patterns reflecting geochemically distinctive lithological units

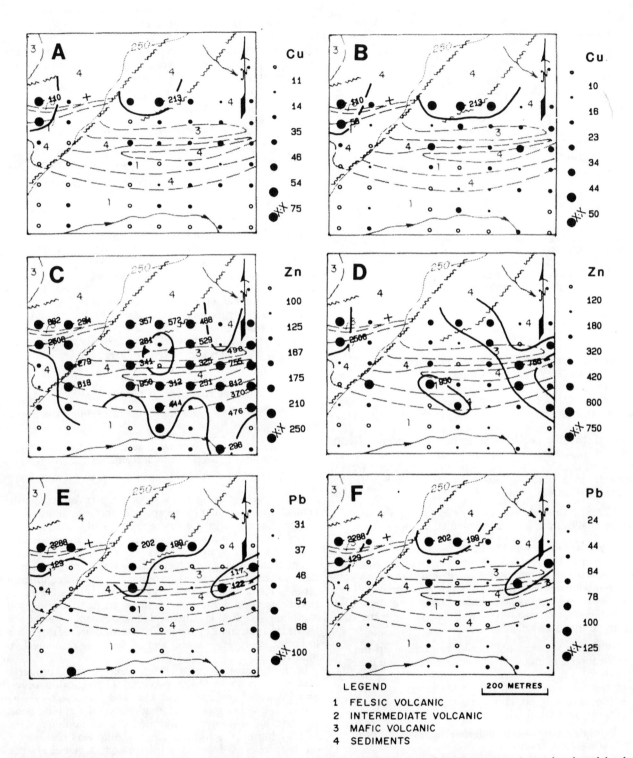

FIGURE 3.25—Soil survey results for Cu, Zn and Pb (Figures 3.25A, C and E, respectively) using contour intervals selected for data from the survey of Figure 3.13 which lies immediately to the northwest. Contour levels were recalculated (Figures 3.25B, D and F) as this portion of the grid was sampled by an experienced sampler. Note the striking differences in metal distribution patterns. All values are in ppm.

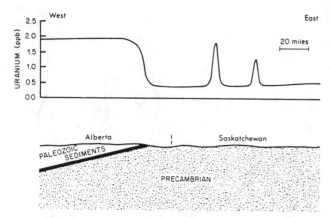

FIGURE 3.26—Diagram illustrating a change in the regional background of U in surface waters on different rock types. Based on analyses of lake and river waters in northern Saskatchewan and Alberta, Canada. Two significant anomalies over the Precambrian rocks have U contents similar to regional background values for waters associated with the Paleozoic sediments. From Levinson (1980).

are typically broad with relatively homogeneous metal contents—Figure 3.26 shows an example for U in waters, but the same principles apply to soils. In contrast, geochemical anomalies related to a mineral occurrence are more likely to appear as sharp spikes as seen on the right of Figure 3.26.

The distinctive geochemical signatures associated with different geological units can sometimes be used in interpretation. Consider the distribution of Cu in soils from a property in a glaciated environment of central British Columbia (Figure 3.27A).

> Can you predict bedrock sources for the Cu soil anomalies from available information?

There are several possibilities and the answer, in general, would be no. The regional topography suggests glaciation from west to east or from east to west along the main valley. Overburden examined during sampling comprises till of a fairly local origin, probably within 500 m of source. Outwash deposits are also recognized, suggesting some complexity to the glacial deposits.

> With this information, would you follow up the Cu anomalies and what procedures would you use?

Sample pulps were reanalyzed for 30 elements by ICP spectroscopy following an aqua regia digestion. Figure 3.27B shows distribution of Ni.

> Does this map help in determining followup procedures?

The answer is very definitely yes. Ultramafic units mapped in the field geologically and geophysically have a strong Ni signature that is displaced about 150 m east of the ultramafic intrusions.

> Can you now predict likely bedrock source areas for anomalous Cu concentrations?

The value of the information provided by Ni greatly exceeds its acquisition cost. Multielement data to aid in this type of interpretation are provided economically by multielement ICP analysis.

Contamination (Column 67)

Figure 3.28 shows results of a humus geochemical survey for Hg in the province of Quebec, Canada (Beaumier, 1983).

> Can you explain the genesis of the Hg anomalies?

The close relationships between roads and anomalies suggests that contamination is probably the source of the Hg—particularly as the easternmost road is the main access to a base metal mine. There is, however, another possibility. In this part of Quebec roads follow eskers to avoid construction problems associated with muskeg on the low lying silts and clays of the Abitibi Clay Belt. The eskers are a distinctive sand and gravel-rich overburden type that might, because of its greater permeability, give higher background Hg values in humus. In either case, the high Hg values associated with the roads would have to be discounted before bona fide anomalies can be defined over the remainder of the survey area.

Coarse fragments (Column 68–69)

It takes very little time and effort to examine the coarse fraction of a soil.

> What importance would you give to this and would you bother collecting this information? Consider the following examples.

A soil geochemical survey was conducted in an area partly characterized by a flat landscape and partly by hummocky terrain. Sampling in the first environment was relatively easy with samples consisting of sandy overburden containing about 5% rounded pebbles. Sampling in the second environment was more difficult. Each hole contained abundant subangular fragments ranging from pea size to large blocks. These descriptions suggest two contrasting overburden environments that could be mapped during soil sampling. The first is a fluvial deltaic deposit, whereas the second is a glacial till environment. Recognition of this leads to better interpretation and should promote cost-effective followup.

In another area, within a known mining camp, a property was acquired for its Pb–Zn potential. Geological mapping was frustrated by lack of outcrop, and a soil survey was chosen to evaluate ground. During geological mapping along soil grid lines, numerous rounded granite boulders were noted lying on the surface. Since granite was not known on the property, these were interpreted as glacial erratics and as evidence for presence of transported till or outwash. In contrast to this interpretation, soil samplers reported that their pits contained about 95% angular fragments that could be fitted together like a jig-saw puzzle. At one site, white coatings on rock fragment surfaces were tested using "zinc-

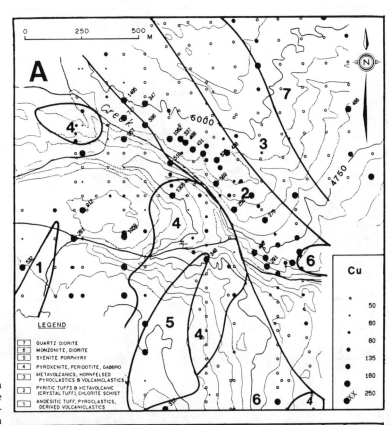
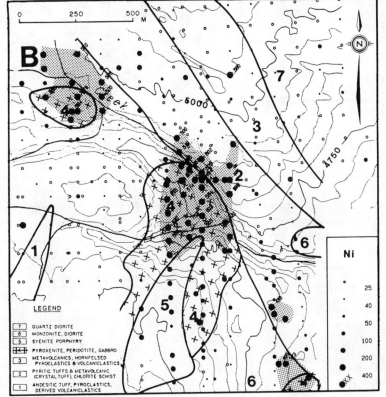

FIGURE 3.27—(A) Distribution of Cu in soils on a base metal property. Can you predict the probable bedrock sources of the anomalous Cu values? (B) Distribution of Ni in the same samples; note the position of the Ni anomaly (shaded) relative to the ultramafic intrusions. Does this information aid your interpretation of the Cu data? All values in ppm.

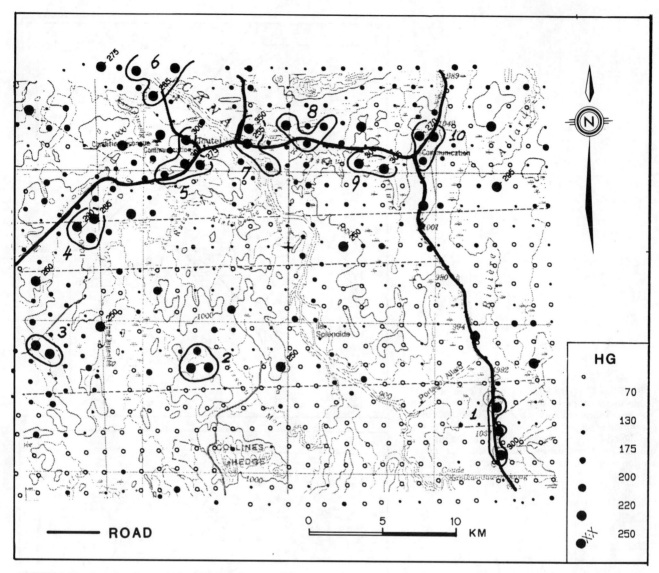

FIGURE 3.28—A regional survey showing distribution of Hg (ppb) in humus samples. Can you explain the genesis of the Hg anomalies?

TABLE 3.7—Time estimates in minutes for soil sample collection with high quality sampling.

	Sample interval					
	25 m		50 m		100 m	
Task	High efficiency	Low efficiency	High efficiency	Low efficiency	High efficiency	Low efficiency
Sampling	1	5	1	5	1	5
Note taking	1	1	1	1	1	1
Traversing	1.5	3	3	6	5	9
Overhead (labelling, bags; load/unload pack)	2	2	2	2	2	2
Total time	5.5	11	7	14	9	17
Samples/hour	10	6	8	5	6	3
Samples/8 hour day	80	48	64	40	48	24

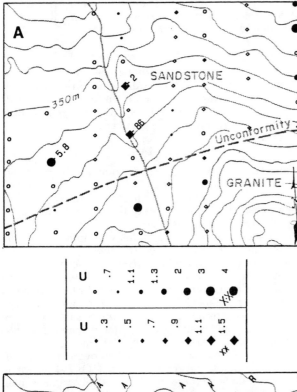
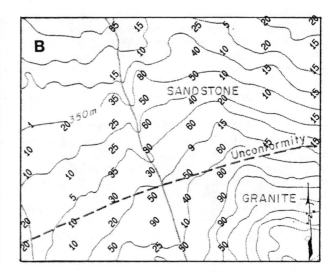
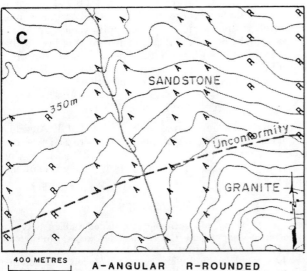

FIGURE 3.29—(A) Distribution of U (ppm) in soils; (B) % coarse fragments; and, (C) shape of coarse fragments for a geochemical survey in the central Northwest Territories, Canada. Data represented by diamonds and circles are from frost boils and normal soils, respectively. The 86 ppm U anomaly occurs in a soil with a very high content (90%) of coarse angular fragments. This information suggests that a local bedrock source for the U can be reliably predicted.

zap," a field test for Zn-bearing minerals, and proved positive. Clearly, the overburden was virtually residual. The rounded granite boulders were glacial erratics that had been deposited immediately on top of the bedrock surface during the Pleistocene, but any fines associated with them had subsequently been washed away. Since glaciation, weathering had produced residual soils up to 0.5 m deep supporting lodgepole pine and alder vegetation. In this case, not only did the geochemical survey benefit from recognition of residual overburden, but property geology could also be mapped in a more reliable way using material from the pits.

The final example is illustrated by a U distribution map (Figure 3.29A) for an area in which bedrock was thought to be covered by thick overburden. The percentage and shapes of coarse fragments in the soils are shown in Figures 3.29B and 3.29C, respectively.

How would you interpret these results? What followup would you recommend?

In this case a combination of favorable geology (sandstone vs. granite) and a high percentage of coarse angular fragments led to the interpretation that the 86 ppm U anomaly was probably close to its bedrock source. Hand trenching subsequently resulted in the discovery of a U occurrence in the underlying bedrock. Other U anomalies in the area lack favorable geology and/or overburden characteristics.

Gamma Count at Sample Site (Column 72–75)

The U exploration boom of the late 1970's introduced many geologists to the scintillometer. The hand-held, total count (eU+eTh+eK) instrument was used primarily to search for radioactivity related to mineralization, but its application for mapping geology was also recognized. Overburden mutes responses, but rock type changes are apparent on traverses. Units such as rhyolite dikes, some shales, granitic intrusions and other lithologies enriched or depleted in one or more of the three radioactive nuclides can be differentiated if their background differences are sufficiently great. Areas of potassic metasomatism, such as in porphyry systems or in the envelope of gold mineralization, can often be mapped in this way.

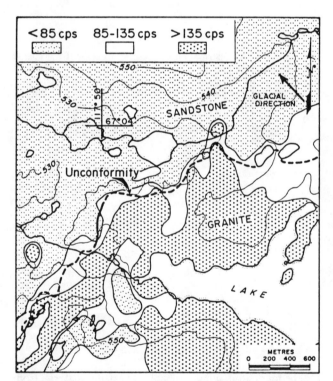

FIGURE 3.30—Example of a ground radiometric survey being used to map geology in an area of extensive overburden cover, central Northwest Territories, Canada (from Hoffman, 1983).

Figure 3.30 illustrates a case where an unconformity between sandstone and granite basement was accurately predicted by a radiometric survey despite an absence of outcrop in an area of very locally derived till and/or residuum. Not shown, but also readily apparent on the field plot, was the ability of the survey to subdivide the granite into high and low background units.

Other Parameters—Composition and/or Site

Notes can be recorded in free format below the coded portion of the geochemical form. Information such as unusual sample appearance, location of cultural features, outcrop areas, creeks, lakes and swamps can be described together with any interpretive comments.

SUMMARY

Many field observations can and should be made in conjunction with soil sampling. Relatively straightforward observations by a trained and interested sampler facilitate a rapid, cost effective appraisal of the mineral potential of the property under investigation.

How much more does high quality sampling cost compared to the more routine "speed sampling"? Let us now examine this question having arrived at an appreciation of the importance of the sampling process. Estimates can be made of the time required to fulfill each of the tasks involved in soil sampling (Table 3.7). Note taking by an experienced sampler can constitute up to 18% of the onsite work in an easy sampling and walking environment where up to 80 samples might be collected per day. The same activity takes about 6% of the total survey time when overburden is difficult to penetrate or traversing is difficult. Overall sampling speed can thus be reduced by 25% to 50% compared to an uncontrolled "quick and dirty" survey in which up to 150 samples might be collected per man day. It is hoped that the reader will have been convinced by the examples throughout this chapter that note taking and high quality sampling are ultimately more cost effective.

ACKNOWLEDGEMENTS

Special thanks are given to Dr. N. Radford of BP Australia for critically reviewing this chapter. Thanks are also extended to Dr. L. Lavkulich of the Department of Soil Science, University of British Columbia, for his assistance with soil taxonomy.

REFERENCES

Beaumier, M. 1983. Pedogeochimie, Region de Brouillan. Services. Geochimie–Geophysique. Government de Quebec, Ministere de l'Energie et des Resources, DP 83–10.

Bradshaw, P.M.D. (editor) 1975. Conceptual Models in Exploration Geochemistry—the Canadian Cordillera and Canadian Shield. Journal Geochemical Exploration, v. 4, 211 pp.

Bradshaw, P.M.D., Clews, D.R. and Walker, J.L. 1979. Exploration Geochemistry. Barringer Research Ltd., Toronto, 54 pp.

Canada Soil Survey Committee. 1978. The Canadian System of Soil Classification. Canada Department of Agriculture Publication 1646, Supply and Services Canada, Ottawa, 164 pp.

Carson, D.J.T., Jambor, J.L., Ogryzlo, P.L. and Richards, T.A. 1976. Bell Copper: geology, geochemistry and genesis of a supergene-enriched, biotitized porphyry copper deposit with a superimposed phyllic zone. In: Sutherland–Brown, A. (editor), Porphyry Deposits of the Canadian Cordillera. Canadian Institute Mining and Metallurgy Special Volume 15, p. 245–263.

Hoffman, S.J. 1983. Geochemical exploration for unconformity type uranium deposits in permafrost terrain, Hornby Bay Basin,

Northwest Territories, Canada. Journal Geochemical Exploration, v. 19, p. 11–32.

Hoffman, S.J. 1985. Soil Surveys for Quebec? Learning from Others. In: La Geochimie d'exploration au Quebec, Seminaire d'Information 1985. Government de Quebec, Ministere de l'Energie et des Resources, DV 85–11, p. 73–81.

Knauer, J.D. 1975. Bell Copper (Newman), British Columbia. In: Bradshaw, P.M.D. (editor), Conceptual Models in Exploration Geochemistry—the Canadian Cordillera and the Canadian Shield. Journal Geochemical Exploration, v. 4, p. 53–56.

Levinson, A.A. 1980. Introduction to Exploration Geochemistry, Second Edition. Applied Publishing, 924 pp.

Soil Survey Staff of the Soil Conservation Service. 1951. Soil Survey Manual. United States Department of Agriculture Handbook Number 18, 503 pp.

Soil Survey Staff of the Soil Conservation Service. 1975. Soil Taxonomy. United States Department of Agriculture, Agriculture Handbook Number 436. United States Government Printing Office, 754 pp.

Travis, R.B. 1955. Classification of Rocks. Colorado School of Mines Quarterly, v. 50, 98 pp.

Worobec, H. and Needham, K. 1981. Canada Mines Handbook 1981–1982. Northern Miner Press Limited, 254 pp.

APPENDIX I
THE CODING FORM
GENERAL INFORMATION—COLUMNS 1–38

Summary

Column	Entry & codes
1–2	SAMPLE TYPE
	40 Bog—upper 100 cm
	42 Bog—below 100 cm
	43 Bog—organic material at mineral horizon interface
	44 Bog—mineral horizon
	50 Soil—top of the B horizon (or top of the horizon if B horizon absent)
	51 Soil—other horizons (organic-rich samples or two samples taken from the same hole)
	52 Frost boil or seepage boil
	55 Deep overburden sample
	58 Heavy mineral concentrate
	60 Talus fines
	68 Heavy mineral concentrate
	70 Biogeochemical sample
	75 Radon, mercury: soil gas
	*90 Special sample—specify and clearly label if high grade
	99 Standard sample
3–4	YEAR
5–7	PROJECT NUMBER
8	PROPERTY OR IDENTIFICATION CODE
9	DUPLICATE SAMPLES
	Label duplicates as 1, 2, etc.
	(collect 1 duplicate pair/30)
10–12	SAMPLER IDENTIFICATION
13–15	SAMPLE NUMBER
19–24	EAST COORDINATE
25–31	NORTH COORDINATE
34–38	NTS MAP SHEET

Example: record 92F/3 as 92F03

Description

Column	Code/Description
1–2	SAMPLE TYPE—Fourteen categories are recognized and each must be identified in the field. Two columns are provided to distinguish different types of "soil" sample, which in effect represent different types of geochemical survey.
	BOGS
	40 Bog—an organic-rich sample from a swampy area
	42 Deep bog—requires a coring device to collect organic material at greater than 1 m depth.
	43 Deep bog organic horizon—above the mineral horizon interface
	44 Bog mineral horizon—requires a coring device to collect below the organic accumulation. If the depth to the mineral sample is less than 50 cm, call the sample a type 50.
	SOILS
	50 Soils—routine survey soil sample
	51 Atypical sample—when two samples are taken from the same hole
	52 Arctic frost boil or seepage boils—in temperate climates
	55 Deep overburden sample—sampling is usually conducted by a contractor. Sample depths are noted and the soil coding form completed where feasible.
	58 Heavy mineral concentrate
	TALUS FINES
	60 Talus fines—fine material collected between angular talus blocks
	68 Heavy mineral concentrate
	OTHER
	70 Biogeochemical sample
	75 Radon or mercury in soil gas
	90 Special samples taken to determine the maximum geochemical response to a known mineral showing. Clearly mark these samples 'high grade'; use fluorescent orange spray paint to identify the anomalous sample bag to avoid carry over contamination at the laboratory.
	99 Standard sample included to check accuracy of laboratory.
3–4	YEAR of sample collection (last 2 digits)
5–7	PROJECT CODE—A list of projects is prepared prior to the field season and amended as new projects arise.
8	PROPERTY OR AREA IDENTIFICATION CODE—A list of properties is prepared prior to the field season and amended as required. The property code identification is used to assist computer processing of the geochemical data. A separate code identifier for orientation studies or detailed work

9   DUPLICATE SAMPLES—Duplicate soil samples (two sample numbers) are collected from holes 1 to 5 metres apart, every 30 stations. Both samples are marked in column 9, the first indicated with a code 1, the second with code 2.

10–12   SAMPLER IDENTIFICATION—Each sampler is assigned a 3-digit number. Assessment of the quality of sampling or note taking by an individual can then be evaluated. Coding errors can be corrected if a sampler systematically misinterprets instructions.

13–15   SAMPLE NUMBER—a 3-digit number beginning with 001. *All blank spaces must be filled with zeros.* Sample numbers are recorded in an ascending sequence. Prenumbering of sample bags is avoided to prevent placing samples in the wrong bag. Labelling of bags on site is subject to duplication errors, but these can often be corrected at the end of the day if an adequate description of the sample is available. The sequence of samples held in the plastic bag used to keep traverse packs dry might also serve as a guide to the correct order. Placing samples on a string at time of collection is recommended. If duplication remains a serious problem, samplers could prenumber five bags at a time corresponding to numbers on one page of the 5 sample computer form. The laboratory should return duplicate numbered bags to the project manager *without preparation* for corrective measures.

19–24   EAST COORDINATE—*right justified*. UTM coordinates completely fill the boxes. Property grid coordinates do not. If a property coordinate is 175E, record as 17500 with *1 blank space* in front of the 1. If the coordinate is 168+13E, record as 16813 with *1 blank space* in front of the 1. If the coordinate is 8+15E, then the number is recorded as 815 with *3 blanks* in front of the 8.

25–31   NORTH COORDINATE—*right justified*. Instructions for recording the north coordinate are similar to those described for the east coordinate.

33–38   MAP SHEET NUMBER—The map sheet number should be entered into the same columns for any property or project. Procedures might be necessary to ensure changes in map sheet label length, from 5 digits to 6 digits, for example, do not affect computer retrieval of data at a later date.

APPENDIX II
THE CODING FORM
SOIL DESCRIPTIVE PARAMETERS—COLUMNS 40–80

Summary

Column   Entry & codes

40   SITE TOPOGRAPHY
       1 Hill top
       2 Gentle slope
       3 Steep slope > 20'
       4 Base of slope
       5 Valley floor
       6 Depression
       7 Level
       8 Rolling
       9 Bog

41   SAMPLE ENVIRONMENT
       1 Tundra-hummocky
       2 Tundra-swampy
       3 Tundra-dry
       4 Grassland meadows
       5 Peat mounds
       6 Bog in depression
       7 Forest-coniferous
       8 Forest-deciduous
       9 Forest-mixed
       A Alder or willows
       B Cultivated
       C Desert, semi-arid
       D Barren
       E Talus fan
       F Bank soil-stream
       G Bank soil-lake
       H Road cut

42   SITE DRAINAGE
       1 Dry
       2 Moist
       3 Wet
       4 Saturated

43   OVERBURDEN TRANSPORT
       L Local
       E Extensive
       U Unknown
       M Mixed

44   WATER MOVEMENT
       S Seepage

45   OVERBURDEN ORIGIN
       1 Till—angular boulders
       2 Outwash—sandy, rounded boulders
       3 Lake sediment—sand, silt
       4 Alluvium—stream deposit
       5 Peat bog
       6 Colluvium
       7 Lake sediment—clay
       8 Talus
       9 Residual
       A Alluvial fan
       B Boulder field*
       G Gravel*
       *Use only if origin not known.

46   BEDROCK
       M Mineralized
       P Present within 100m upslope
       D Present within 100m downslope
       B Underlies sample site
       G Gossan
       F Fe surface stains
       R Radioactivity

47–48   pH

49	SAMPLE TEXTURE	70	SHAPE OF COARSE FRAGMENTS

49 SAMPLE TEXTURE
    0 Organic muck
    1 Fibrous, peaty organic matter
    2 Very sandy
    3 Sandy
    4 Sand-silt
    5 Sand-silt-clay
    6 Silt
    7 Silt-clay
    8 Clay
    9 Gravel
50–51 THICKNESS OF SOIL SAMPLE INTERVAL (CM)
52–54 BOTTOM OF SOIL SAMPLE INTERVAL (CM)
55–56 SOIL HORIZON
    LH Leaf, humus, undecomposed vegetation on the surface
    AH Dark brown to black, organic-rich mineral horizon, usually not more than 15cm deep
    AE Gray to white (occasionally brown) leached mineral horizon near surface, usually sandy; accompanied by BF or BT horizon at depth
    BH Black, organic-rich mineral horizon at depths greater than 15 cm
    BF Red-brown, Fe-rich horizon
    BT Brown, clay-rich horizon
    BG Water saturated most of the year; red-brown mottles
    BM Brown horizon, only slightly different in appearance from underlying parent material
    C1,C2 . . . etc: Parent materials
    CA White $CaCO_3$ precipitate in C horizon
    01,02 . . . etc: Bog samples at different depths
    TF Talus fines
57 SOIL TYPE
    C Chernozem—prairie soil, usually under grassland or meadow; thick AH (>10cm); CA at depth
    S Solonetz—saline soil, high salt content
    L Luvisol—BT horizon diagnostic
    P Podzol—BF horizon diagnostic
    B Brunisol—BM horizon is only B horizon
    R Regosol—little or no soil development. No B horizon, only LH (sometimes) and C
    G Gleysol—BG horizon diagnostic
    O Organic soil, bog vegetation, no mineral matter
58–60 LOCAL BEDROCK COMPOSITION
    Estimate using lists 1–4
61–66 COLOR
    Munsell notation or abbreviation
67 CONTAMINATION
    Blank if none
    C Culvert
    F Farming
    G Garbage
    H House
    I Industry
    L Logging
    M Mine
    R Road
    T Trench
    O Other—specify
68–69 % COARSE FRAGMENTS

70 SHAPE OF COARSE FRAGMENTS
    A Angular
    R Rounded
    S Subrounded, subangular
    M Mixture of above types
71 SCINTILLOMETER NUMBER
72–75 GAMMA COUNT AT SAMPLE SITE
    Scintillometer reading at ground level over hole
76 ROCK
    *Asterisk if bedrock is influencing scintillometer counts
77–78 APPROXIMATE SLOPE ANGLE
79–80 APPROXIMATE SLOPE DIRECTION

Description

Column Code/Description

40 SITE TOPOGRAPHY—Position of a soil sample on the landscape is critical to geochemical interpretation. Environments represented by each of the codes are subjectively determined.

41 SAMPLE ENVIRONMENT—The character of, or lack of, vegetation at the sample site might assist in relocating anomalous sites and/or identify geobotanical indicators of mineral occurrences. The number of selection possibilities appears formidable, but in any one area there are usually only a few possibilities. Anomalies or apparent anomalies correlating with unusual sample environments, such as bogs in otherwise well drained soil environments, are important for interpretation. Base metal "kill zones" or other geobotanical indications may reflect suboutcropping mineral occurrences where high metal concentrations severely affect plant growth. These should be obvious to the samplers and can assist followup if known in advance (e.g., from scanning air photographs).

42 SITE DRAINAGE—A subjective indication of degree of water saturation of the soil is recorded. Soils are normally in a moist condition (code 2) except where the climate is semiarid (code 1) or perhaps in more temperate climates, when it has not rained for a long time. Soils with code 1 must be moistened prior to a color determination. Wet soils (code 3) characterize seepage areas. Saturated soils (code 4) characterize seepage zones and areas below the water table.

43 OVERBURDEN TRANSPORT—Overburden typically covers 95% to 99% of the landscape. Direction of transport of surficial deposits and/or a prediction of the location of a bedrock source for the overburden are obviously important to interpreting the geochemistry. These can be estimated by trained soil samplers on site. The soil sampler looks at the upper 50 to 100 cm of the overburden routinely and can offer an opinion on overburden origin. An educated guess is based on knowledge of local bedrock geology (from published maps or by examining outcrops). Boulder types, their angularity, concentration and variety, add to the

interpretation. Many angular blocks, all apparently of the same rock type fitting together in a three dimensional 'jig-saw' puzzle, suggests a local (code L) source, whereas a jumble of rounded boulders or exotic rock types suggest extensive (code E) transport. Geochemical anomalies in the first case would likely suggest a mechanical genesis whereas in the second case a hydromorphic genesis might be suspected. An interpretation of the probable source of the overburden has to be made at the time of sampling; the quality of the estimate will depend on the training and experience of the sampler.

44 DIRECTION OF WATER MOVEMENT—Water normally percolates downwards through the soil profile.

Water movement in base of slope zones and seepages is upwards. Seepage zones (code S) are specifically monitored for geochemical patterns related to hydromorphic dispersion of metals from a concealed deposit upslope. Upwelling water in seepage areas and evaporation or evapotranspiration can promote metal accumulation, even in areas not associated with a mineral occurrence. Many seepage zones are located at subtle topographic inflections, and it would be unwise to rely solely on topographic or airphoto interpretation, in place of field observations, to predict their existence.

45 OVERBURDEN TYPE—Overburden type, like overburden transport, can be identified relatively easily by the trained observer. Much time and effort are not generally required. The sampler need only examine available road or creek cuts and on digging the soil hole note obvious physical properties. The sampler is not assumed to have the skill of a Quarternary geologist. The overburden classification must be relatively simple. Detailed classifications might be required under some circumstances, such as for differentiating between units of the same type of overburden (i.e. on deep overburden surveys conducted in glacial terrain of the Canadian Shield), and experts should be sought to establish these classifications where needed.

1 Till—Nonsorted, nonstratified glacial overburden deposited by a glacier; contains angular and/or striated boulders; typically more clay-rich than outwash.
2 Outwash—drift deposited by meltwater beyond the margin of glacier ice; stratified; contains many rounded boulders. Typically more sand-rich than till.
3 Lake sediment, sand—former beach or deltaic lake deposit.
4 Alluvium—sediment deposited by streams. To be suspected in valley bottoms beside active creek channelways.
5 Peat—dark-brown or black organic accumulation exceeding a thickness of 0.5 metres, produced by the partial decomposition of mosses, sedges, trees, and other plants growing in bogs, swamps, marshes and wet places.
6 Colluvium—overburden moving downslope.
7 Lake sediment, clay—former lake bottom sediment, also includes deposits of varved clay.
8 Talus—broken rock fragments and fines accumulating at the foot of a steep slope or cliff.
9 Residual—overburden derived from bedrock disintegration in situ. Downslope dispersion is minimal.
A Alluvial fan—unsorted debris deposited by streams in front of a mountain range in response to a decrease in flow gradient.
B Boulder field
G Gravel

Note that options 6, B and G are used only when the initial origin of the overburden cannot be determined. If the overburden today is colluvium, it might be possible to guess its previous genesis, for example from the texture of the fines and the nature of the boulders till might be recognized: the till designation, code 1, should then be recorded instead of the colluvium code 6.

46 BEDROCK OUTCROP—Knowledge of the position of bedrock outcrops can assist interpretation of geochemical results. Geological control offers a mechanism for rating the relative importance of anomalies. Geological mapping normally involves traversing from outcrop to outcrop based on recognition of their location on airphotos. Only a fraction of a property might be traversed systematically by a geologist on a line by line basis analogous to collecting soil samples. This being the case soil samplers can be asked to note the presence of outcrop and predict areas where bedrock immediately underlies the soil pit (code B).

M Mineralized—contains some form of economic minerals.
P Bedrock *present* within 100 metres *upslope*
D Bedrock *present* within 100 metres *downslope*
B Bedrock underlies the sample location
G Gossan—iron oxide residue remaining after chemical weathering of sulphide minerals.
F Iron oxide surface stains—looks like a Gossan, but leaching has not destroyed the character of the bedrock.
R Radioactivity—bedrock has above average or anomalous radioactivity

47–48 SOIL pH affects mobility of many elements, such as U, Mo, Cu and Zn (Table 3.3). The following procedure, suitable for use in a field camp, can be used to determine soil pH:

1. Place 1 level tablespoon (15 ml) of minus 10-mesh soil in a 100 ml Dixie cup.
2. Add 2 tablespoons (30 ml) of deionized water. Stream water may be suitable if its pH is neutral (6.5 to 7.0) and its conductivity is low (less than 20 umhos).
3. Mix the soil and water with a coffee stir stick.
4. Mix the suspension 3 more times at 10 minute intervals.
5. Allow the suspension to settle for 30 minutes.
6. Measure pH with pH meter and record the value in columns 47 and 48. Note the decimal point

is already printed on the coding form. If pH is greater than 10.0, record as 99 and specify in note section of form. The pH electrode must be entirely within the supernatant liquid and must be gently agitated during measurement to obtain a stable reading. If a pH meter is not available, decant the liquid into a clean test tube, allow the suspension to settle (1 hour), add 2–3 drops of pH indicator liquid, and compare the solution color to a color chart provided with the liquid indicator.

49 SOIL TEXTURE—Sample texture is determined in a approximate fashion by hand analysis. A portion of the sample is placed between thumb and forefinger and a rotating, rubbing motion is initiated. Sand grains are immediately evident; silt is perceived as a soapy feeling; clay as a sticky substance which can be rolled and flexed like plasticene. Organic matter is divided into two categories: well decomposed muck (*code 0*) and recognizable debris such as leaves, twigs and needles (*code 1*). Every soil sample probably contains a little organic matter, sand, silt, clay and pebbles. Column 49 attempts to identify the major constituents of the sample and classify the sample accordingly.

50–51 THICKNESS OF SAMPLED INTERVAL—in cm (2 boxes), right justified, no decimal point. If thickness is 100 cm or greater, report 99 and record the correct sample thickness on the note section of the form. Soil samples are normally taken over an interval of 10 cm: difficulty in sampling should be suspected if sample interval is less than 5 cm. This information is omitted for deep overburden samples which assumes continuous sampling from one depth to the next.

52–54 BOTTOM OF THE SAMPLE—in cm, right justified, no decimal point. For *deep overburden samples* (sample type 55), depth of the sample is recorded in metres in columns 51 to 54, *with a decimal point in column 53*.
Depth of the sample checks the accuracy of the horizon designation of columns 55 and 56. For example, if the sample is called an AE horizon, and the sample is taken at 60 to 75 cm depths, the horizon identification is probably incorrect.

55–56 SOIL HORIZON—Nomenclature can be very involved so the computer codes used are simple abbreviations of the main horizons to avoid continual reference to a computer code key. In the absence of orientation studies, soils are normally collected from the top of the 'B' horizon (i.e. BF—iron-rich, BT—clay-rich, BG—iron oxide mottles or, BM). If the 'B' horizon is not developed, sample material is taken from the 'C' horizon. Organic material (LH, AH, BH) and leached 'A' horizon material (AE) are usually avoided.

57 SOIL TYPE—For routine sampling programs, soil type is not an important descriptor, being automatically defined once the B soil horizon to be sampled has been identified. Soil type is required for orientation surveys or where several horizons or non-diagnostic horizons are sampled.

58–60 ROCK TYPE—LOCAL BEDROCK COMPOSITION (See Appendix III). The ability to record information about geology depends on sampler skills and on availability of outcrop. Geological information can be added to geochemical records after geological interpretation. Geology coding formats can range from phoenetic abbreviations (i.e. granite recorded as GRT), to numeric codes. The following classification is provided, based on rock nomenclature from Travis (1955).

A three column code is used and depending on the knowledge of the sampler, either 1, 2, or 3 columns are filled. The first code is determined after identifying the rock type to be either intrusive (code I), volcanic (V) sedimentary (S), or metamorphic (M). The second code defines the gross composition of the rock. For intrusive rocks categories listed in list 1 of Appendix III are used. Coding nomenclature for volcanic, sedimentary and metamorphic rocks follow list 2, 3 and 4 of Appendix III, respectively. The third code describes the rock type.

Because the code is built in three stages, there are no blank spaces at the beginning (col. 58) although there can be at the end. Examples of rock type codes illustrate the system.

Rock type	List 1	List 2	List 3	Code
Hornblendite	Intrusive(I)	Ultrabasic(4)	Hornblendite(3)	I43
Syenite	Intrusive(I)	Intermediate(2)	Syenite(1)	I21
Andesite	Volcanic(V)	Andesite(2)		V2
Limestone	Sedimentary(S)	Calcareous(3)	Limestone(1)	S31
Gneiss	Metamorphic(M)	Gneiss(7)		M7
Amphibolite	Metamorphic(M)	Phaneritic(2)	Amphibolite(7)	M27

61–66 SOIL COLOR—can be determined using a soil color chart prepared by the Munsell Company. The chart is used by pedologists. Alternatively, colors can be estimated without reference to color charts, at the expense of introducing systematic differences between samplers. Soil color reflects organic matter content (blacks, grays), leaching, (light colors, white), Fe (red-brown), Mn (blacks) and geology (variable imprints). Color changes might define geological contacts and under favourable conditions, provide evidence of a suboutcropping mineral occurrence.

If the Munsell chart is unavailable; easily remembered abbreviations, filling up to 6 columns of information, are suggested—the soil must be moist:

L—light	M—medium	D—dark
OR—orange	RE—red	YE—yellow
PI—pink	GR—green	BL—blue
PU—purple	BR—brown	BK—black
GY—gray	WH—white	RB—red brown
OB—orange brown		

Generally the first two letters of the color are used, except for GY, BK, RB, and OB. Flexibility is allowed to avoid the need to "look-up" predefined codes.

67 CONTAMINATION—Contamination can be introduced from two sources: at the sample site and by the sampler. Site contamination is usually

obvious to the sampler, or to the followup crew if contamination has not been reported adequately. Contamination is most likely on old properties or in mining camps where ground has been disturbed 50 to 100 or more years ago and has since returned to a near "virgin-appearing" state. Sampling of old roads, staging areas, or dumps, can provide significant geochemical reponses.

The second source of contamination is the sampler. His sampling device must be free of contaminants, particularly if low sample weights are collected. Chrome plated trowels, for example, can contribute significant Mo in addition to Cr contaminants. High grade specimens carried in the sampler's pockets, can introduce contamination if the sampler put his hands in his pockets, particularly if they are wet. Contaminants adhering to the sampling device from the preceeding station(s) must be avoided, particularly if the last station was at a dump or workings of an old mine or prospect. Do not use a shovel that has been previously used on known prospects, particularly if Au exploration is involved.

"Carry-over" contamination of moist, unmineralized soils is unavoidable and cleaning the sampling device of material from the preceeding pit is not usually practical. It is therefore imperative that material placed in the soil bag not have come in contact with the sampling blade. This is easily accomplished by digging a sufficiently large clod of dirt and ensuring material showing evidence of the sampling device (i.e. shovel marks), is not included in the sample.

68–69 % COARSE FRAGMENTS—The percentage of coarse fragments exceeding a pea size dimension (including boulders) in the soil sample or pit is estimated by volume. An exact measurement is not possible nor needed, but the sampler should be able to differentiate sites with a very high proportion of coarse material, such as in talus fans, from sites containing no boulders or pebbles at all, such as varved clays.

Coarse fraction information provides data on overburden composition and on proximity of the site to bedrock. Changes in stone contents from one soil hole to the next could be an indication of an overburden change. Site specific variability and systematic differences between samplers, or for the same sampler from day to day, have to be discounted to define regional patterns.

70 SHAPE OF FRAGMENTS—Average degree of angularity of the coarse fragments, together with their percentage from columns 68 and 69, gives an impression of type and degree of transport of the overburden. If boulders of one rock type are angular (code A) and present in large quantities bedrock is probably exposed nearby. Conversely, overburden containing rounded pebbles and cobbles (code R) suggests outwash or alluvium. Such information may assist mapping of overburden types and/or suggest followup methodologies. For example, prospecting of boulders left beside an anomalous soil hole or trench might be recommended if a high proportion of angular material was noted.

71 SCINTILLOMETER NUMBER CODE—A total count scintillometer normally accompanies U exploration programs, but could be used on any routine survey to assist in mapping geology, alteration or overburden. Scintillometers are expensive devices not normally retired from service when new models become available. Radiometric surveys on one property might employ several instruments and a code specific for each unit is recorded to enable corrections of systematic variations if necessary.

72–75 SCINTILLOMETER COUNT—right justified, *no leading zeros*. Scintillometer readings are taken at ground level over the soil pit. Note is made of landscape geometry which abnormally enhances or depresses scintillometer readings. The sampler can test the effect geometry of the ground relative to the scintillometer has on a reading by moving the scintillometer from above the hole to the bottom of the hole. Scintillometer readings over outcrops are recorded in the notes so that overburden counts can be compared to bedrock values.

76 ROCK—High concentrations of boulders or outcrop underlying or adjacent to the sample site can enhance a background scintillometer reading up to 20% to 30%, sufficient to classify a background reading as anomalous. The asterisk (code *) brings attention to these situations.

77–78 SLOPE ANGLE—right justified in degrees. Only an approximation is necessary.

79–80 SLOPE DIRECTION—an approximate indication of slope direction (codes—S, SE, SW, N, NE, NW, E, W) helps predict the likely upslope source of metals.

APPENDIX III
THE CODING FORM
ROCK TYPE CLASSIFICATION—COLUMNS 58–60

LIST 1 INTRUSIVE ROCKS

I1	QUARTZ RICH
I11	Granite
I12	Quartz Monzonite
I13	Granodiorite
I14	Quartz diorite
I2	INTERMEDIATE
I21	Syenite
I22	Monzonite
I23	Diorite
I24	Gabbro
I3	FELDSPATHOID RICH
I31	Nepheline syenite
I32	Nepheline monzonite
I40	ULTRABASIC
I50	CARBONATITES
I60	SPECIAL TYPES
I61	Pegmatite

I62 Aplite
I63 Lamprophyre
I64 Trap
I65 Felsite
I66 Intrusion breccia
I67 Diabase

LIST 2 VOLCANIC ROCKS

V0 UNDIFFERENTIATED*
V1 BASALT*
V2 ANDESITE*
V3 DACITE*
V4 RHYOLITE*
V5 QUARTZ LATITE*
V6 LATITE*
V7 TRACHYTE*
V8 PHONOLITE*
V9 NEPHELINE LATITE*
V21 *Fine grained flow
V22 *Porphyritic flow
V23 *Crystal tuff
V24 *Ash tuff
V25 *Lapilli tuff
V26 *Agglomerate
V27 *Lapilli breccia
V28 *Block breccia
V29 *Turbidite
  *Column 59 can be "1" through "9", "2" is used as an example.

LIST 3 SEDIMENTARY ROCKS

S1 ARENACEOUS
S11 Siltstone
S12 Mudstone
S13 Greywacke
S14 Sandstone
S15 Quartzite
S16 Conglomerate
S2 ARGILLACEOUS
S21 Shale
S22 Argillite
S3 CALCAREOUS
S31 Limestone
S32 Dolomite
S4 CHEMICAL PRECIPITATE
S41 Chert
S42 Marble
S43 Iron formation

LIST 4 METAMORPHIC ROCKS

M1 FINE GRAINED CONTACT
M2 PHANERITIC
M21 Meta quartzite
M22 Marble
M23 Soapstone
M24 Hornfels
M25 Serpentine
M26 Skarn
M27 Amphibolite
M28 Eclogite
M3 MECHANICAL
M31 Mylonite
M32 Flaser
M33 Augen
M34 Ultramylonite
M40 SLATE
M50 PHYLLITE
M60 SCHIST
M7 GNEISS*
M8 MIGMATITE*
M81 *Granite
M82 *Monzonite
M83 *Granodiorite
M84 *Conglomerate
M85 *Sandstone
M86 *Augen
M87 *Granulite
M88 *Quartz diorite
M89 *Diorite
M80 *Amphibolite
  *Column 59 can be "7" or "8", "8" has been used as an example.

Chapter 4

ANALYSIS OF SOIL SAMPLES

W. K. Fletcher

INTRODUCTION

Analysis of soil samples involves a series of activities (Table 4.1) that can be grouped into sample preparation, sample dissolution (though this is omitted with x-ray fluorescence and direct neutron activation analysis) and the final analytical determination (Figure 4.1). Together these activities define the *analytical system*. In addition, the quality and reliability of the data must be monitored throughout the analysis.

The preliminary steps in the analysis, i.e., sample preparation and dissolution, can drastically influence the results obtained and their suitability for the intended purpose (Table 4.2). Conversely, the final analytical step largely governs detection limits (or analytical sensitivity), the type and severity of interference problems likely to be encountered with different sample types and the number of elements that can be determined simultaneously. The last is, of course, an important factor in analytical costs and we are fortunate that advances over the last three decades, first from colorimetry to atomic absorption (AAS) and now to the inductively coupled plasma (ICP), have maintained remarkably steady costs while expanding capabilities for multielement determinations.

Possibly because of the ease of obtaining analytical data and increasing reliance on sophisticated instrumentation in centralized (often commercial) laboratories, there continues

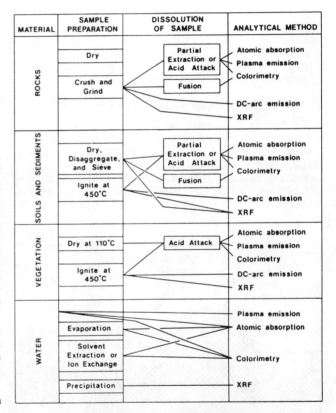

FIGURE 4.1—Some of the pathways for preparation, dissolution and analysis of exploration samples. After Fletcher (1981).

to be a major communications gap between the exploration geologist and analyst. Too often this leads to blind adherence to standard methods (such as minus 80 mesh; hot acid digestion; atomic absorption base metals; ICP multielement, etcetera), which may be quite inappropriate, and to interpretation of analytical artifacts as significant geological trends. In contrast, choice of appropriate analytical procedures should be considered a key factor in the design of a well organized survey and one that will facilitate, not complicate, anomaly recognition and interpretation.

In this section we will consider both the choices and

TABLE 4.1—Summary of activities in a geochemical laboratory in usual sequence.

1. Receipt of samples from client
2. Allocation of work
3. Sample preparation
4. Sample weighing
   - insertion of laboratory control standards
5. Sample dissolution or leaching
6. Element determination
   - calibration of instrument
   - determination of element
   - dilution and rerunning of above range solutions
   - quality control evaluation
7. Repeat analysis if required
8. Collation of results and transfer to computer
9. Transfer results to client
10. Archive pulps and rejects

TABLE 4.2—Comparison of six stream-sediment analyses for Ni (from Hansuld et al., 1969).

	Preparation		Digestion	Ni (ppm)
	crushing	fraction		
Lab A	no	minus 80	70% HClO$_4$	20
Lab B	no	minus 80	1:3 HNO$_3$	60
Lab C	no	minus 80	HNO$_3$/HCl	150
Lab D	no	minus 80	1:1 HCl	320
Lab E	yes	minus 100	1:1 HCl	14
Lab F	no	minus 100	1:1 HCl	1120

Sample description
stream sediment with 0.5% magnetite
magnetite contains an average of 0.28% Ni
80% of magnetite is minus 80 mesh
99% of sample is plus 100 mesh and 96% is plus 80 mesh

opportunities provided by the analytical system and methods of monitoring its reliability. Before doing so, however, a review of the distribution and behavior of trace elements in soils and related media is in order.

DISTRIBUTION OF TRACE METALS IN SOILS

Weathering decomposes fresh bedrock to give, depending on its duration and intensity, soils consisting of as yet undecomposed primary minerals (principally quartz and feldspar) mixed with clay minerals and oxides of iron and alumina (Figure 4.2). Texturally the primary minerals usually predominate in the coarser silt and sand fractions, whereas weathering products are most abundant in the finer fractions. The biosphere adds organic compounds, principally plant litter in various stages of decay and humic substances, to the surface layers of this mixture. In addition, soils derived from mineral prospects may contain primary ore minerals or their secondary alteration products. Any trace metal is therefore usually present within a soil in several distinct forms, notably (Fletcher, 1981):

(1) In lattices of undecomposed primary minerals.
(2) In lattices of secondary minerals or occluded in amorphous compounds, for example, in the lattices of clays or in amorphous or crystalline iron oxides.
(3) Associated with organic matter either from uptake by the living organism or by complexation and chelation by organic compounds in soils.
(4) Adsorbed on surfaces of clays, iron and manganese oxides and organic matter.
(5) As major constituents of surviving ore minerals, e.g., Sn in cassiterite.
(6) As major constituents of secondary products derived from ore minerals, e.g., Cu as malachite.

Speciation (i.e., the distribution of elements between different constituents or components of a soil) of metals within soils is relevant to exploration geochemistry insofar as the analytical system should differentiate metal derived from mineralization from that representing the background component. This maximizes anomaly contrast while minimizing other sources of variability. The ideal of anomalous metal patterns derived only from mineralization and superimposed on an otherwise flat background can seldom be realized. Information on speciation of metals therefore serves a secondary purpose—to identify and classify geochemical patterns related to mineralization and distinguish them from those resulting from other processes. In either case successful application of the analytical system requires some geochemical or mineralogical insight into the form in which metal derived from mineralization is most likely to be incorporated in the soil. Two broad generalizations follow from our consideration of the forms of metals present in soil:

(1) Metals present in primary silicates or ore minerals (either as part of the lattice or encapsulated as inclusions) will often be found in the coarser size fractions (*free* grains of micron Au would be an exception), and a strong decomposition procedure will probably be required to release the metal.
(2) Metal adsorbed on surfaces of clays, Fe and Mn oxides or organic matter, possibly having been carried to its present site in solution, will be in the finer fractions of the soil and extractable with a weak or partial decomposition technique.

These generalizations correspond, respectively, to the extremes of mechanical (clastic) and hydromorphic (chemical) dispersion. Many anomalies or dispersion patterns, however, will contain components of both.

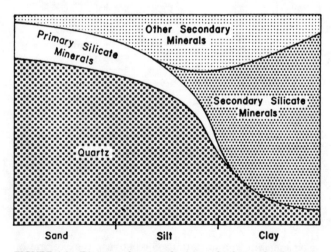

FIGURE 4.2—Diagram showing the general relationship between particle size and kinds of minerals present. Quartz dominates the sand and coarse silt fractions. Primary silicates such as feldspars, hornblende, and micas are present in the sands but tend to disappear as one moves to the silt fraction. Secondary silicates dominate the fine colloidal clay. Other secondary minerals such as the oxides of iron and aluminum are prominent in the fine silt and coarse clay fractions. (After Brady (1974). Reprinted with permission from "The Nature and Properties of Soils", 8th edition. (C) Macmillan Publishing Co., Inc., 1974.)

SAMPLE PREPARATION

Preparation of soils usually involves drying and disaggregation followed by sieving to obtain the desired size fraction (Table 4.3). Choice of size fraction is the most important consideration. However, before discussing this in detail some minor problems that might be encountered should be noted:

(1) Contamination from equipment wear (use nylon rather than brass or silver-soldered stainless steel screens) or sample carryover.
(2) Loss of Hg (and possibly other volatiles) if drying exceeds 65°C (Koksoy et al., 1967). Too high a temperature also bakes clays making subsequent disaggregation difficult.

The optimum size fraction for analysis should, of course, initially be chosen from the results of an orientation survey to determine which fraction provides good contrast most reliably. In addition, if best contrast is found to be in the coarser fractions, this advantage must be balanced against the greater costs involved if it becomes necessary to grind samples to ensure representivity of subsamples and obtain adequate analytical precision. For elements occurring in the primary mineralization in unstable minerals (e.g., sulfides), which are at least partly decomposed during weathering with redistribution of their metal content throughout the (finer) components of the soil, the minus 80 mesh fraction (<177μ) is often a useful compromise that provides adequate precision without the need to grind the sample prior to analysis. However, some examples of situations where use of minus 80 mesh material may be less than optimum or unacceptable include:

TABLE 4.3—Sieve sizes for American Society for Testing Material (ASTM) meshes.

Sieve number	Nominal aperture μm	Sieve number	Nominal aperture μm
10	2000	80	180
18	1000	100	150
20	850	120	125
40	420	140	106
45	355	170	90
50	300	200	75
60	250	230	63
70	212	270	53

(1) The metal is present only in coarse grains of a resistate mineral or is encapsulated within undecomposed rock or mineral fragments that require crushing and grinding prior to decomposition. Examples of this include occurrence of Sn as coarse grains of cassiterite (Figure 4.3) and Au in undecomposed rock fragments (Figure 4.4).

(2) Oxidation of sulfides has redistributed metals to very fine fractions. In these circumstances, anomaly contrast is improved by using fractions appreciably finer than 80 mesh. Shilts (1984) provides an excellent example of this in going from unoxidized to oxidized tills (Figure 4.5).

(3) Metals redistributed during weathering by hydromorphic processes may be concentrated

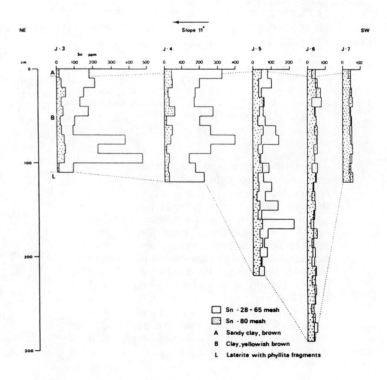

FIGURE 4.3—Comparison of distribution of Sn as cassiterite in minus 80 mesh and minus 28 + 65 mesh soils associated with tin mineralization in Malaysia. Note that only the coarser size fraction shows increasing Sn values in the anomalous zone intersected by pits J-3 and J-4. Modified from Thanawut et al. (in press).

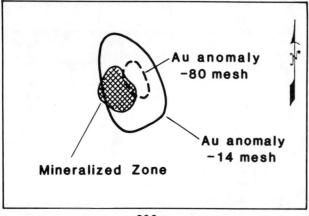

FIGURE 4.4—Comparison of the extent of Au anomalies in minus 80 and minus 14 mesh soils associated with the Saddler Prospect, Lander County, Nevada. Modified from Wargo and Powers (1978).

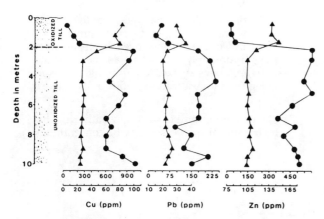

FIGURE 4.5—Influence of oxidation on the Cu, Pb and Zn content of ("●" and upper scale) sand size heavy minerals SG > 3.3; and, ("▲" and lower scale) clay size fraction fractions of glacial till in southeastern Quebec. Note that the concentration scales for the two fractions are not the same. After Shilts (1984).

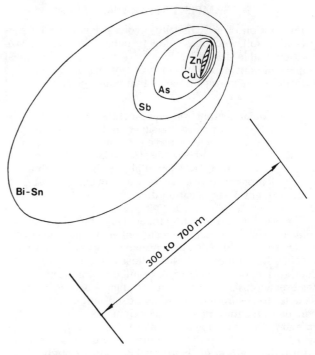

FIGURE 4.6—Conceptual plan view of a pisolitic laterite halo about a postulated buried gossan. The halo is shown elongated in the direction of wash that prevailed during its formation. After Smith et al. (1979).

in the coarser fractions of the overburden. This can occur if sand grains and gravels acquire a coating of secondary iron oxides that scavenge the trace metals from solution. DiLabio (1985) has attributed increased abundance of gold in very coarse fractions of oxidized till to this process. In this situation crushing would cause dilution, and greatest anomaly contrast would probably be obtained by selective dissolution of the coatings as described by Filipek et al (1982) for stream sediments. In a very different environment lateritic pisolites, in which metals accumulate with concretionary iron oxides, have been found to provide a suitable exploration medium at Golden Grove, Australia (Figure 4.6).

SAMPLE DECOMPOSITION

Introduction

If the element of interest is present in a mineral that resists conventional decomposition techniques and a total element determination is required, it may be preferable to analyze a solid sample by x-ray fluorescence or direct neutron activation. The latter has the additional advantage that it is nondestructive, and samples of interest can subsequently be examined mineralogically (once the induced radioactivity has fallen to safe levels). However, both atomic absorption and plasma emission spectroscopy, the principle methods of analysis for exploration samples, require that the elements to be determined be in solution. This introduces the additional step of sample decomposition into the analytical scheme but gives the geochemist considerable flexibility to liberate only that portion of the metal related to the mineralization sought. Conversely, use of inappropriate decomposition techniques can mask significant anomalies and highlight irrelevant geochemical patterns related to other processes. The decompositions and leaches used are divided into *strong and partial* (or weak) in Table 4.4, these terms being preferred to their less appropriate synonyms *total* and *cold* extractions, respectively.

Strong Decompositions

Of the strong decompositions, digestions with acid mixtures are preferred to fusions (except for fire assay, which

TABLE 4.4—Summary of decomposition techniques (based on Fletcher, 1981).

Decomposition	Reagents
Strong decompositions	
1. With hot concentrated mineral acids	nitric, hydrochloric & perchloric acid mixtures
2. Fusions:	
acid	potassium bisulfate
alkaline	sodium carbonate; sodium hydroxide; lithium tetraborate with or without oxidizing agent
reductive	fire assay
3. Sublimations	ammonium iodide for Sn
Partial decompositions	
1. Nonselective	cold, dilute hydrochloric acid; buffer solutions; EDTA
2. Selective:	
for organic matter	hydrogen peroxide; sodium hypochlorite
for Fe and Mn oxides	hydroxylamine hydrochloride; ammonium oxalate; sodium dithionite
for sulfides	potassium chlorate-hydrochloric acid; hydrogen peroxide; bromine

is a specialized reductive fusion) because of the tendency of the latter to produce solutions having high salt contents that clog instrument burners and nebulizers. Mixtures of nitric:hydrochloric or nitric:perchloric are widely used with a 0.2–1.0 g sample being weighed into a test tube and then leached by the hot acid solution for a specified time. It must be emphasized that these are not total extractions and that the proportion of the total metal content liberated varies considerably as a function of sample mineralogy. Resulting geochemical patterns can therefore be expected to reflect the lithologies of different soil parent materials, which will have their own characteristic backgrounds. Furthermore, even with similar acid mixtures, extraction efficiency will vary with the sample:solution ratio and duration and temperature of the extraction. Since these operational procedures inevitably vary somewhat between laboratories it is not advisable to change analysts part way through a program.

Strong decompositions are often used routinely on geochemical samples where their failure to provide a total extraction is not regarded as a disadvantage since . . . . "the significant trace amounts of metal in exploration have quite likely been introduced into the rocks by hydrothermal or other genetic processes, and such metals are easily solubilized by boiling nitric acid. Background amounts of metals such as copper and zinc in crystal lattices of silicates are less significant in exploration, and the need to solubilize them is not as important in exploration as in abundance and distribution studies." (Ward et al., 1969). Thus the main requirement of the decomposition is to liberate that fraction of the metal derived from potentially interesting prospects. Problems will be encountered with acid decompositions when they are used in situations where they are incapable of achieving this goal. Three distinct possibilities can be envisaged:

(1) The metal is associated with a resistate phase and is inaccessible to the decomposition technique; e.g., Cr as chromite is not decomposed by hot aqua regia (Table 4.5) and Sn as cassiterite as described in the case history at the end of the chapter.
(2) Metal derived from mineralization is only present in soils in a relatively weakly retained form and the strong acid decomposition gives either (a) unnecessarily complex geochemical pat-

TABLE 4.5—Variation in extraction of Cr from different lithologies with five digestion procedures (I. Thomson, personal communication).

Lithology	Chromium extracted (ppm)				
	A*	B	C	D	E
Chromite	11.00%	930	600	140	nd
Pyroxenite	2900	1600	1550	480	nd
Chlorite schist	5800	1350	1400	390	nd
Serpentinite	2500	2200	2050	740	nd
Quartzite	270	115	42	18	nd

*A = alkaline fusion; B = hot aqua regia; C = hot nitric acid; D = hot 0.5N hydrochloric acid; E = cold 0.25% EDTA; nd = not detected

terns related to both mineralization and lithological variations, or (b) a weak anomaly is swamped out by variations in high background values associated with silicates and other resistate minerals. The latter situation is likely to be particularly true for anomalies resulting from hydromorphic transport of metals to seepage sites or breaks in slope.

(3) Metal is liberated from the anomalous phase by the decomposition procedure, but coprecipitation or some other process results in its loss from the analytical system. Loss of Pb as lead sulfate after aqua regia digestion of samples with high sulfide contents is an example.

Dolezal et al (1968) provide an extremely useful guide to the effects of various reagents on different minerals. Some important minerals that will not readily succumb to strong acids are summarized in Table 4.6 with suggested decomposition procedures. In these circumstances, however, it may be preferable to use an instrumental method, such as x-ray fluorescence or direct neutron activation, which avoids the need for preparation of a solution by determining total metal content of a solid sample. This is, of course, often the situation in dealing with heavy mineral concentrates.

Determination of Au is a rather special case insofar as the only mineral acid able to completely dissolve it is the mixture of concentrated nitric and hydrochloric acids known as aqua regia. This decomposition is often used, in conjunction with solvent extraction, for determination of gold by atomic absorption. Organic matter is not digested and, if present in significant amounts, must be removed by a preliminary ignition of the sample at 500–600°C. Fine Au fully encapsulated in silicates is also likely to go unreported. Similar considerations also apply to the use of hydrobromic acid-bromine or cyanide solutions to dissolve Au. Fire assay and, more recently, direct neutron activation analysis may therefore be preferred methods of analysis depending on the mode of occurrence of the Au.

With respect to fire assay, a half assay ton [(14.6 g) or assay ton (29.167 g)—the factor used in North America so that 1 mg of gold per assay ton is equivalent to one troy ounce in one avoirdupois ton (2000 lb); in the UK and Australia the long ton (2240 lb) is used and the assay ton is 32.667 g] portion of the sample is first fused at 1000°C with a flux consisting of variable proportions of litharge (lead oxide), a silver collector, sodium carbonate, sodium borate, potassium nitrate, flour or charcoal to ensure reducing conditions and silica. The molten mass is then transferred to a mold where Au collects with silver in the lead button that forms below the silicate slag. The solidified button is physically separated from the slag and transferred to a bone ash cupel, which is placed in a second muffle furnace at 800°C. Here Pb is absorbed by the cupel to leave a precious metal bead.

In classical fire assay gold is parted from the silver with nitric acid and the undissolved gold weighed (gravimetric finish). However, if detection limits as low as 1–5 ppb are required, Au content of the bead can be determined by neutron activation or, after aqua regia dissolution, by atomic absorption or plasma emission spectroscopy. Overall reliability of fire assay is very dependent on the experience and skill of the assayer and in particular the choice of flux composition. Haffty et al (1977) give a much more detailed overview of the steps involved and problems associated with the determination of Au and other noble metals by fire assay.

Partial Extractions

Partial extractions can be classified as either selective or nonselective depending on their ability to release metals from particular phases of the soil (Table 4.4). For example, hydroxylamine hydrochloride, which is selective, can be used to release metals from secondary manganese oxides; whereas dilute hydrochloric acid, which is nonselective, will simultaneously liberate metals from several phases. Though capable of providing valuable information on geochemical dispersion processes, use of selective partial extractions, particularly in sequential extraction schemes (Fletcher, 1981), is a specialized endeavor that should normally be entrusted to a geochemist.

Nonselective partial extractions are commonly used when hydromorphic anomalies (which may be the only expression of concealed mineralization) are the target. Under these circumstances a weak extraction, only capable of removing

TABLE 4.6—Decomposition techniques for some resistant minerals. The presence of significant concentrations of an element as a constituent of these minerals will cause relatively low results to be reported if routine geochemical decomposition techniques are used.

Mineral	Decomposition techniques	
	Fusion	Acid extraction
Barite	Na_2CO_3	—
Beryl	Na_2CO_3; NaOH	—
Cassiterite	NH_4I; Na_2O_2	—
Chromite	NaOH; Na_2O_2	$HClO_4$
Columbite/tantalite	NaOH; Na_2O_2	HF
Fluorite	Na_2CO_3; NaOH	Be nitrate solution
Gold	Fire assay	Aqua regia; bromine; alkaline cyanide solutions
Monazite	Na_2O_2	H_2SO_4; $HClO_4$; H_3PO_4
Zircon	NaOH; Na_2O_2; borax	

metals associated with amorphous phases or adsorbed on surfaces, will provide better contrast than a strong decomposition and at the same time depress patterns related to lithological variations. Similarly, extraction of a large proportion of the anomalous metal content of a soil by a partial extraction will often indicate presence of a hydromorphic anomaly (Figure 4.7). Plots of the ratio of partial:strong extractable metal are therefore useful interpretive guides in anomaly evaluation.

Cold dilute hydrochloric acid (~0.5–1M) is probably the most widely used partial extractant. It will release exchangeable and adsorbed metals, dissolve carbonates and partly release metals from oxide phases, organic matter and such secondary minerals as plumbojarosite. Sulfides can also be decomposed: caution must therefore be exercised in interpreting anomalous patterns in unoxidized glacial tills as hydromorphic as they might also result from mechanical dispersion of sulfides. In soils with varying carbonate contents its buffering action on solution acidity can influence extraction efficiency. Sufficient acid should be used to avoid this problem.

Other partial extractants include the pH buffer solutions used in colorimetric field tests (e.g., ammonium citrate–hydroxylamine hydrochloride); ethylenediaminetetraacetic acid (EDTA), a chelating agent with some selectivity for organic-bound metals; and acidified solutions of hydroxylamine hydrochloride if the metal is suspected of being associated with secondary manganese oxides (Chao, 1972).

ANALYTICAL METHOD

The analytical method used to estimate metal concentrations is of less geochemical significance than the preceeding steps providing it is sufficiently sensitive to reliably measure the concentration sought and is free from serious interference problems. Ensuring that these conditions are met at a reasonable cost is largely the responsibility of the analyst. Nevertheless, it is useful for the geologist to have a general appreciation of the capabilities of the principal analytical methods and the elements they are most suited to determining at the concentrations found in geological materials. These are summarized in Figures 4.8, 4.9, 4.10 and Table 4.7.

The inductively coupled plasma, with its high throughput, wide dynamic range and ability to provide multielement data with precision comparable to that of atomic absorption, enables data for a wide range of major and trace elements to be obtained at low cost (Thompson and Walsh, 1983). The additional information provided can be a useful interpretive guide to changing geological or geochemical conditions on a soil grid. However, it must be remembered that the analytical results are a function of the sample preparation and the decomposition procedures—rather than the ICP itself. Thus they will have little or no significance if the element of interest is present in a mineral not attacked by the decomposition. In this context, although strong acid decompositions give less than total extractions for the major element constituents of silicate minerals, it may still be possible to detect geologically significant, and hence useful, patterns.

Each of the analytical methods has a concentration range over which its response to increasing concentrations is essentially linear and also its own particular interference problems. Because of the very variable bulk composition of geological materials and their extremely wide range of trace element concentrations, it is in the customer's own interest to identify to the laboratory samples having either unusual compositions or very high metal contents. Assay grade samples should certainly be flagged or submitted separately whenever possible: this will reduce the number of dilutions required to bring a high grade sample on scale, and help avoid interference problems and carryover contamination. Interference problems caused by changes in composition of samples may, of course, still occur. The geologist should therefore always consider if changes in metal concentrations or associations are geologically or geochemically reasonable.

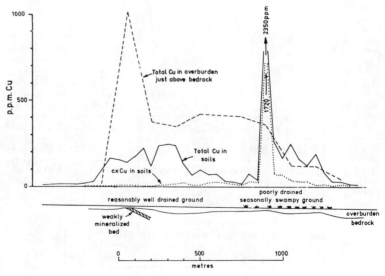

FIGURE 4.7—Example of hydromorphic copper anomaly associated with a residual anomaly over a copper occurrence in central Zambia. After Reedman (1979).

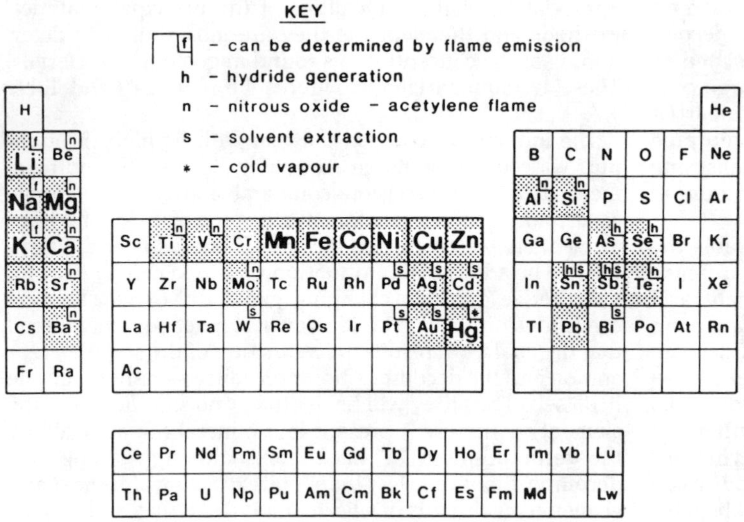

FIGURE 4.8—Analysis of exploration samples by flame atomic absorption. Elements most suitable for determination shown in stipple; bold face letters indicate that the concentrations of the elements can normally be estimated without difficulty after sample decomposition with strong acids. Small letters indicate that special operating conditions are required or recommended as shown in the key. After Fletcher (1981).

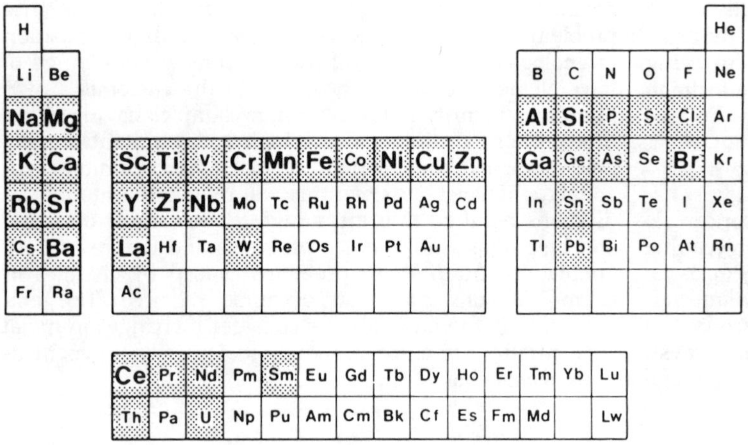

FIGURE 4.9—Analysis of exploration samples by x-ray fluorescence. Concentrations of elements in bold face and stipple can be estimated in most samples. Concentrations of elements in stipple only are close to or below detection limits, they will only be measured in samples with above average contents. After Fletcher (1981).

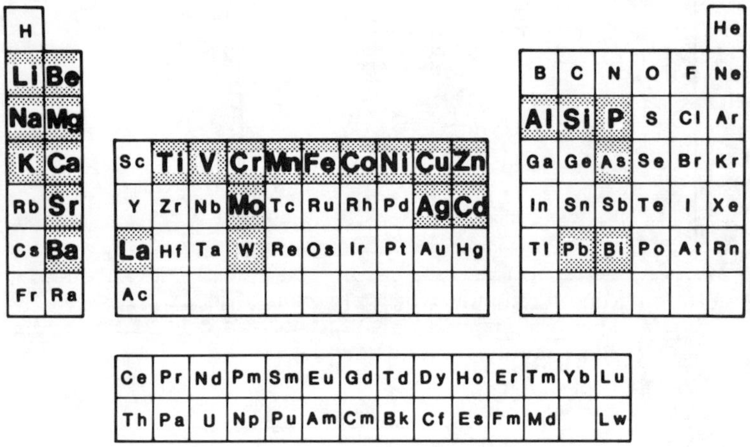

FIGURE 4.10—Analysis of exploration samples by inductively coupled plasma. Concentrations of elements in bold face and stipple can be estimated in most samples without difficulty following sample decomposition with strong acids. Concentrations of elements shown in stipple only are close to or below detection limits in many samples or present special problems. For elements, such as Ba and Cr, that are likely to be present in minerals not decomposed by strong acids results will be considerably less than total values—the data may, however, still be very useful. Based on data provided by Chemex Labs, Vancouver and Thompson and Walsh (1983).

TABLE 4.7—Comparison of some characteristics of analytical methods.

Method*	Capital $	Multi-element	Determinations per day	Comments
Colorimetry	1×10^3	No	20–100	very simple; adaptable to field
AAS	2×10^4	No	500	easy to set-up; several elements can be determined on same solution but not simultaneously
ICP	$1–3 \times 10^5$	Yes	>2000	needs skilled analyst to supervise; computer; sample in solution
XRF	$1–5 \times 10^5$	Yes	>1000	needs skilled analyst to supervise; computer; analyzes solid sample

*AAS = Atomic absorption spectrophotometry
ICP = Inductively coupled plasma spectrometry
XRF = X-ray fluorescence

QUALITY CONTROL AND RELIABILITY

For most elements a modern analytical laboratory is capable of providing data beyond the normal requirements of an exploration geochemical programme with respect to detection limits, reproducibility and accuracy. However, to request analyses of the highest possible quality would be inordinately time consuming and costly—particularly when it is realized that much of the variability encountered in exploration geochemical data arises not in the laboratory but from the natural variability of the material sampled or from sampling errors. Consequently, in exchange for high productivity and low costs, the exploration geochemist accepts some loss of analytical quality. This does not, however, mean that quality and quality control are simply discarded. In fact it becomes even more important to ensure that errors are monitored and maintained within acceptable limits. Much of this task is the responsibility of the analyst. However, insofar as the analyst is seldom involved in the end use and has no information on the sources or magnitude of other errors, the geologist or geochemist must ultimately accept (or reject) the suitability of the data for their intended purpose. Fortunately, this can be done at little extra cost as part of the routine analytical program as described by Fletcher (1981).

Errors in the analysis can be random or nonrandom if positive or negative bias is systematically introduced (Figure 4.11). Random errors (noise) cause poor precision whereas systematic errors, which affect relative accuracy, are associated with contamination or changes (drift) in the analytical system. Precision is of concern because decisions in exploration geochemistry often involve the interpretation of relative differences in element concentrations and these might be masked by excessive noise (Figure 4.12). Systematic changes in the analytical system are of concern as a source of false geochemical patterns. Monitoring of random errors is described first.

Random Errors and Precision

We will begin by assuming that random errors are normally distributed with a bell-shaped distribution around their mean (Figure 5.3). Exploration geochemists then usually define analytical precision (P_c), for concentration c, as the percent relative variation at the 95% (two standard deviations) confidence level:

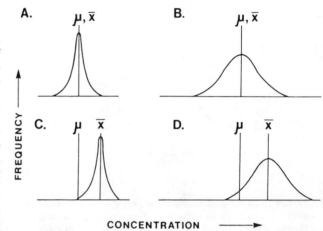

FIGURE 4.11—Random and systematic errors. The variation in concentration caused by random errors is represented by the normal curve with average value X: μ is the true concentration of the analyte. A: the dispersion (width) of the normal curve is small and symmetrical around μ, i.e. X = μ and results are both accurate and precise. B: the dispersion is greater than in A but still symmetrical about μ—precision is therefore relatively poor and although the average value (X) is accurate, this is not necessarily true of individual determinations. C: dispersion is narrow but a systematic positive error has been introduced (X > μ)—results are precise but inaccurate. D: systematic error and poor precision. After Fletcher (1981).

$$P_c (\%) = 200 \cdot S_c/c$$

where S_c is an estimate of the standard deviation at concentration c. A precision of ±10% for a Cu concentration of 100 ppm indicates that 95 out of 100 analyses would lie between 90 and 110 ppm. Precision of ±10–15% is generally regarded as acceptable for most exploration purposes. The detection limit for an analytical system is defined as the concentration at which precision becomes ±100%. It follows that an acceptable precision can only be achieved at concentrations several times higher than the detection limit (Figure 4.13).

The standard deviation and precision could be estimated by replicate analyses of special control standards at several

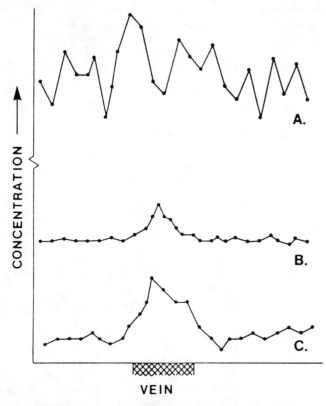

FIGURE 4.12—Influence of analytical precision on anomaly contrast. Noisy data (A) arising from random errors in sampling and/or analysis obscures the anomaly over the mineralization. Although absolute concentrations are lower in (B) and (C), the data are less noisy and anomaly contrast improved. After Fletcher (1981).

concentrations. This, however, would be counter productive and might lack relevance to precision achieved with real samples if bulk composition or textural differences exist between samples and standards. Thompson and Howarth (1978) have described a more elegant method in which precision is estimated on the basis of duplicate analyses of actual samples. In this procedure the average of duplicate analyses ($[X_1+X_2]/2$) estimates true concentration, and the absolute difference ($|X_1-X_2|$) estimates the standard deviation at that concentration. Steps in the procedure are given in Table 4.8 with a worked example in Table 4.9 and Figure 4.14. A minimum of 50 duplicate analyses are recommended: this can easily be achieved by *randomly* selecting and reanalyzing 5% or fewer of the samples from most soil grids. Calculations are readily programmed for a microcomputer.

The foregoing discussion assumes that random errors are normally distributed. However, if a trace element is present almost entirely as a major constituent of a few very rare mineral grains, for example flakes of gold or tin in cassiterite, the probability of finding no such grains in a sample

TABLE 4.8—Procedure for estimation of precision by the Thompson and Howarth (1978) method.

1. Randomly select samples and resubmit them for analysis using a new identification number.
2. When a minimum of 50 pairs of duplicate analyses have been obtained, calculate their means ($[X_1+X_2]/2$) and absolute differences $|X_1-X_2|$.
3. Arrange the list in increasing order of concentration means.
4. From the first 11 results obtain the mean concentration and the median difference.
5. Repeat 4 for each successive group of 11 samples ignoring any remainder less than 11.
6. Calculate or obtain graphically the linear regression of the median difference on the means.
7. From the regression the value of $|X_1-X_2|$ can be taken as the standard deviation (s) at concentration X_c. P_c is then calculated as $P_c=200s/X_c$.

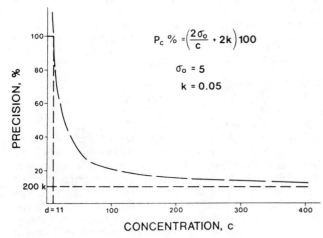

FIGURE 4.13—Variation of precision with concentration. The detection limit (in this case = 11 ppm) is the concentration corresponding to a precision of ± 100%. After Fletcher (1981).

or subsample increases as the grains become rarer or sample size decreases. This probability can be estimated with the Poisson distribution from the relationship:

$$P_n = e^{-\hat{z}} \cdot \hat{z}^n/n!$$

where P_n is the probability of finding n grains if $\hat{z}$ is the average number of special grains in a sample of the specified size. An example is given in Table 4.10 and Figure 4.15. If $\hat{z} < 5$, results are strongly skewed towards reporting low values until, with $\hat{z} < 1$, many replicate analyses report only background concentrations accompanied by erratic (spot) highs.

The standard deviation of the Poisson distribution is equal to the $\sqrt{z}$ so that the relative error (RE) is given by:

$$RE\ (\%) = \sqrt{z}/z = 1/\sqrt{z}$$

TABLE 4.9—Estimation of precision by the Thompson and Howarth (1978) method. In this table part of a paired duplicate data set have been arranged in ascending order and divided into groups of 11; means and median differences for each group are obtained and plotted in Figure 4.14.

Pair number	$(X_1+X_2)/2$		$|X_1-X_2|$
1	3.7		1.1
2	4.0		0.2
3	4.0		0.6
4	4.6		0.8
5	4.8	Group mean = 5.6	0.2
6	5.0	Group median = 1.1	2.0
7	6.3		1.6
8	6.5		1.5
9	7.1		2.2
10	7.3		0.3
11	8.5		1.4
12	9.3		1.7
13	11.7		3.2
14	12.2		1.6
15	13.1		2.2
16	18.1	Group mean = 19.6	2.3
17	18.1	Group median = 2.3	4.1
18	20.1		2.2
19	24.6		3.2
20	27.1		1.8
21	29.1		4.2
22	32.0		4.0
23	33.7		4.5
24	35.7		4.2
25	38.7		5.4
26	41.7		5.4
27	43.3	Group mean = 51.1	6.3
28	51.6	Group median = 5.4	8.6
29	55.5		0.2
30	58.0		2.0
31	65.5		10.6
32	66.3		4.2
33	71.7		5.5
34 etc.	76.8		9.5

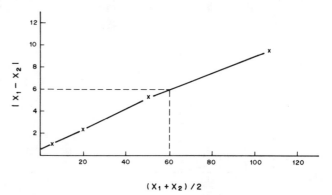

FIGURE 4.14—Regression of median differences ($|X_1-X_2|$) against averages ($[X_1+X_2]/2$) for data in Table 4.9. Precision at a concentration of 60 is equal to ± 20% at the 95% confidence level.

Thus, with 20 special particles in the sample or subsample, as recommended by Clifton et al. (1969) in their classic paper on sampling errors for Au, the relative error becomes approximately ±50%—a barely acceptable value. Every effort should therefore be made at both the sampling and subsampling stage to ensure that this criterion is met either by (i) using larger samples, (ii) concentrating the special grains (e.g., by panning or with heavy liquids), or (iii) grinding to reduce particle size. It should be noted that the last option only works if the number of special grains is increased and that grinding the matrix or gangue material alone is of no value. Grinding can also result in contamination problems as described later.

Systematic Errors

Systematic errors resulting from contamination, drift in the system with time and physical or chemical interferences, caused by variation in sample composition, can introduce false trends into the data (Figure 4.16). Typically slight differences in technique between analysts and espe-

TABLE 4.10—Estimation of the probability of a subsample containing n special grains if the average number of special grains in subsamples of the specified size is 1.

Number of special grains	Calculation	Probability (%)
0	$P_0 = e^{-1} = 0.37$	37.0
1	$P_1 = \dfrac{1 \times 0.37}{1} = 0.37$	37.0
2	$P_2 = \dfrac{1 \times 0.37}{2} = 0.185$	18.5
3	$P_3 = \dfrac{1 \times 0.185}{3} = 0.0625$	6.3

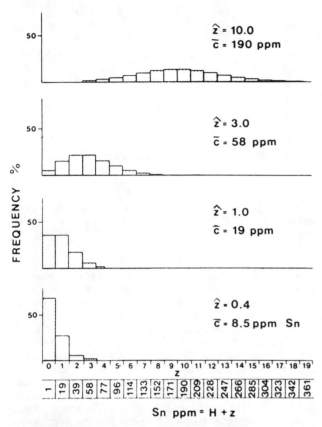

FIGURE 4.15—Histograms of the Poisson distribution with different values for the average number of special grains ($\hat{z}$) in a sub-sample. Corresponding concentrations of Sn have been calculated assuming that a 100 mesh grain of cassiterite contains 18.9μ of Sn and there is 1 ppm of Sn in the gangue. After Fletcher (1981).

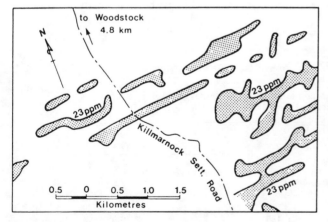

FIGURE 4.16—Spurious Cu anomalies introduced into a soil survey by analytical errors. The east-west orientation of the anomalies parallels the soil traverse lines and has resulted from the analysis of the samples in the same order as their collection. After Fletcher (1981).

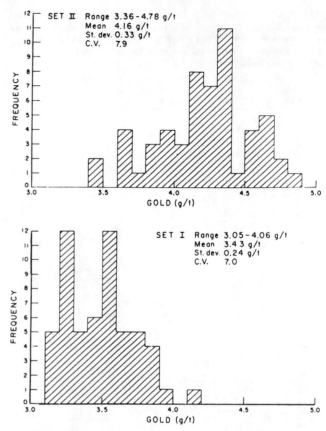

FIGURE 4.17—Comparision of duplicate analysis for Au by two analysts using different intruments. After Allcott and Lakin (1975).

cially between different laboratories, even though they are nominally using the same procedure, result in systematic differences between their results (Figure 4.17). It follows that it is unwise to change laboratories part way through a sampling programme.

Contamination

Contamination can result from wear and tear of equipment during sample collection and preparation (Tables 4.11 and 4.12), from use of impure reagents in decomposition or from corrosion of instrumentation by acidic solutions or organic solvents. However, a more important problem is usually carryover contamination if samples with high values contaminate those that follow (Table 4.13). This is likely to be particularly troublesome when assay grade material is included in batches of background samples: such material should therefore be identified and processed through a separate preparation facility whenever possible (check that your laboratory provides this service). Some mixing of samples with high and low metal contents is, however, inevitable in a (successful) geochemical survey. A series of declining values following a very high value should therefore be viewed with suspicion until verified by analysis of (hopefully uncontaminated) coarse rejects. This, of course, requires the order in which samples have been prepared to be known.

TABLE 4.11—Some sources of contamination from equipment during sample preparation.

Material	Potential contaminants
Steel and iron grinding plates	Fe, Co, Cr, Cu, Mo, Mn, Ni, V
Alumina ceramic plates	Al, Cu, Fe, Ga, Li, Ti, B, Ba, Co, Mn, Zn, Zr
Tungsten carbide	Co, Ti, W
Lubricants	Mo

TABLE 4.12—Contamination of sample with Ni from pulverizer plates (S. J. Hoffman, personal communication).

Sample #	Ni ppm
2716	1470
2717	780
2718	730
2719	850
2720	740

Normal Ni concentration range of samples was 5–20 ppm. A rough correlation between extent of contamination and degree of silicification was observed.

Contamination of soils during sample collection is seldom a problem. However, if humus is to be analyzed, its contamination by inorganic mineral grains can result in extremely erratic results (Table 4.14). Considerable care may be needed during sample collection if this problem is to be avoided.

Drift

Instrumental drift results if the response (sensitivity) of the instrument (e.g., an atomic absorption spectrophotometer) slowly changes with time. This is usually not a serious problem because laboratories routinely recalibrate their instruments throughout the day. Potentially more troublesome systematic changes can occur during sample decomposition if reagent strength, the temperature of the decomposition or the extraction time vary. Operating conditions must therefore be carefully maintained. A rather different form of drift sometimes occurs if there is an interval of several days between sample decomposition and the analysis during which metals are lost from solution by precipitation or adsorption on residues (Table 4.15).

Interferences

New methods of analysis are often heralded (by inventors and manufacturers) for their freedom from interferences. This initial euphoria is eventually displaced by a more realistic assessment of the interferences that inevitably arise from differences between the physical and chemical characteristics of the sample (solutions) and the standard (solutions) used for calibration. Some examples for atomic absorption are summarized in Table 4.16.

In designing an analytical system the analyst will attempt to overcome or minimize interferences associated with the type of material being analyzed. However, exploration samples are particularly troublesome because of their complex character and the very variable compositions likely to be encountered even in the same batch. Obviously the exploration geologist cannot be expected to be familiar with all the interferences that might be encountered; it is therefore sensible to discuss departures from a normal (silicate) compositional range with the laboratory. An example of suppression of high U values by abnormally high Mn contents is shown in Table 2.3. If in doubt, results should be checked by different methods. Differences in absolute concentrations may be acceptable, but are relative trends similar and probably (geologically) valid?

Monitoring Systematic Errors

Checks for carryover contamination require samples to be analyzed in a known sequence. In addition, most laboratories insert control standards in every batch of samples—usually at a frequency of about 1 in 40. Analyses of these controls are plotted to monitor changes in the response of the analytical system (Figure 4.18). This information should be requested from the laboratory, or blind controls (easily prepared from a few bulk samples) should be included in

TABLE 4.13—Carryover contamination of U in a suite of lithogeochemical samples (S. J. Hoffman, personal communication).

Pulverizing sequence	U ppm	Notes
1	260,000	Pitchblende vein
2	850	Nonradioactive dolomite
3	195	Nonradioactive dolomite
4	15	Granite gneiss
5	12	Granite gneiss
6	2	Amphibolite
7	2	Granite gneiss
8	1	Amphibolite
9	<1	Dolomite
10	1	Granite gneiss

TABLE 4.14—Effect of contamination of humus samples by inorganic material in the determination of Au by direct neutron activation analysis (S. J. Hoffman, personal communication).

Sample location	Au ppb	
	Contaminated	Uncontaminated
1	15	1
2	17	1
3	14	2
4	40	2
5	<20	2
6	12	1
7	41	3
8	20	7
9	160	3
10	69	2
11	15	7

TABLE 4.15—Losses of Pb and Ag (ppm) from nitric-perchloric acid digestions (S. J. Hoffman, personal communication).

Sample	Lead		Silver	
	A	B	A	B
2	352	87	17.4	2.9
3	13	11	0.9	0.1
4	91	72	6.2	4.5
5	346	116	18.6	2.8
6	39	35	1.1	0.5
7	37	34	1.5	0.3
8	12	9	0.7	0.2
9	96	74	6.7	4.5
10	98	77	7.3	5.0
11	12	10	1.0	0.2
12	39	38	1.7	0.6
13	367	86	16.4	2.4

A = analysis within 5 hours of digestion; B = analysis after 5 days.

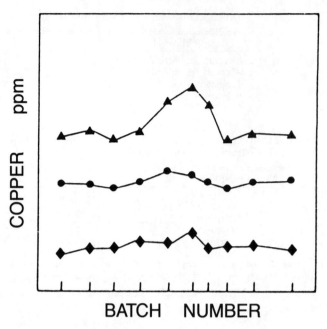

FIGURE 4.18—Control graph based on the analysis of three laboratory standards. After Fletcher (1981).

every sample batch submitted. It is also useful to outline the extent of different analytical batches on geochemical maps and check that this, rather than geology, is not the source of apparent geochemical trends.

SUBMISSION OF SAMPLES FOR ANALYSIS

Commercial laboratories usually offer several combinations of sample preparation and analytical procedures. Choice of the most appropriate combination must consider:

TABLE 4.16—Some interferences in the determination of trace elements in geological matrices by flame atomic absorption. From Fletcher (1981).

Element	Interference[1,2]	Comments
Ag	B	Background correction
Ba*	C, I	Supression by Si, Al and P; enhancement by alkalies and alkali earths
Be*	C	Supression by Al
Cd, Co	B	Background correction
Cr	C	Supression by Fe, Na and K; enhancement by Al, Mg, Ca
Mo*	C	Supression by alkalies, Ca & Fe—add up to 1000 µg/ml Al
Ni, Pb	B	Background correction
Rb	I	Add K ionization buffer
Sr*	C, I	Add K ionization buffer and La
Zr*	C	Add NH_4F

[1] All determinations in air-acetylene flame except (*) in nitrous oxide-acetylene flame
[2] B = background absorption significant; C = chemical interference; I = ionization interference

(1) the geochemical criteria discussed in the sections on sample preparation and decomposition
(2) the amount of sample required
(3) the number of elements to be determined and if this can be done with a single decomposition procedure to minimize costs
(4) if analytical sensitivity is adequate to detect and estimate concentrations at an acceptable precision
(5) freedom from systematic errors
(6) if throughput (turnaround time) is adequate to keep pace with the exploration program—particularly at the height of the season
(7) cost.

Initially this choice may require discussion with one or more laboratories and submission of samples from orientation surveys for analysis by a variety of procedures. Once the analytical system has been defined it is equally important that it be adhered to even if responsibility for submission of samples for analysis changes. At the same time misplaced samples, transposed numbers and other clerical errors must be avoided. To facilitate this most laboratories provide sample submittal forms to accompany each batch of samples.

A typical form requests: total number of samples; number and types of each sample material; size fractions for sieving or grinding; decomposition procedure to be followed; elements to be determined and detection limits required; whether an assay or geochemical analysis is needed; instructions for disposition of rejects (i.e., the unused coarse fractions) and pulps (i.e., unused sieved or ground material prepared for analysis); and project codes and instructions for transmittal of results and billing. In addition, as already noted, it is essential to flag high grade material and advise the laboratory of the presence of unusual samples whenever possible. Many laboratories discard rejects after 30 days and pulps after six months. The costs of additional storage (especially for pulps) are usually minimal compared to those of any resampling that might become necessary if samples are discarded.

AN ANALYTICAL CASE HISTORY: TIN EXPLORATION IN EASTERN NORTH AMERICA

(Contributed by I. Thomson)

This case history comes from personal experience and is based on real data. It has been deliberately disguised to prevent embarrassment to any of the individuals involved in the actual exploration programs. Furthermore, events have been dramatized to make the example more interesting and to emphasize some of the key mistakes.

May we present "A Tale of Two Companies" and introduce first Silver Mountain Allied Resource Trading and Exploration known everywhere as SMART–EX. Smart–Ex acquired what they believed to be a prime position for a tin play by staking the entire outcrop of a small stock, The Tombstone Granite, in which a high grade tin prospect was known (Figure 4.19). In addition they took up a large land holding around the stock in the hope that high grade veins

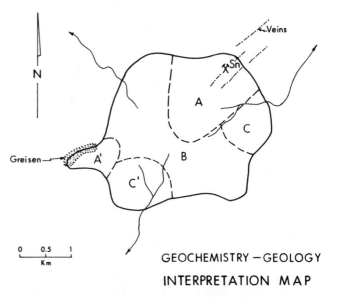

FIGURE 4.19—Geology—geochemistry interpretation map of the Tombstone Granite. A–A1: acid extractable tin phase (biotite/muscovite granite); B: non-cassiterite bearing, total tin in heavy mineral phase (hornblende/sphene granite); and, C–C1: lower tin content phase of B (hornblende/sphene granite).

or, more optimistically, greisen zones might extend into the country rock of mixed Paleozoic sediments.

The known occurrence was a vein trending NE carrying cassiterite with values of up to 3% Sn over 2 m widths along a strike length of 40 m. Some past production had been obtained from an open cut known as the Sardine Tin Mine.

The area is rolling upland with deep cover and very few outcrops. Smart–Ex decided that a soil survey would be the best way to cover this unexplored area. The company thought it wise to use a proven method and selected B Horizon soil sampling. They opted for multielement analysis, to give comprehensive supporting data, and selected a package deal offered by a large commercial laboratory giving twenty six elements simultaneously, including tin, by plasma determination following nitric-perchloric acid digestion.

Samples were duly collected and sent to the lab where, in the absence of any instructions from the geologist, they were sieved to minus 80 mesh prior to analysis: a standard procedure.

> By now several flaws have marked the Smart–Ex project. Pause now and make a list of the problems you can see in the Smart–Ex approach to this exploration program.

The Smart–Ex geologists were impressed by the amount of analytical data they received and, to retain a sense of direction, gave special attention to the tin results (Figure 4.20). In truth they had more numbers than they knew how to deal with, even with computer support. They were rather disappointed in the tin data, noting broad flat patterns of 10–30 ppm tin with no sign of the sharp anomaly peaks they had expected. Worse still, there was no indication that

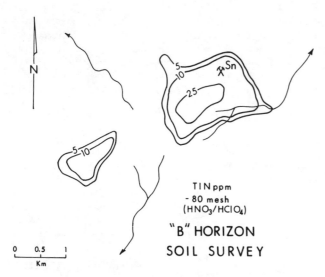

FIGURE 4.20—Tombstone Granite: Sn content (ppm) of minus 80 mesh B horizon soils using a nitric-perchloric acid decomposition.

the geochemistry was "seeing" the mineralization. First impressions were that soil geochemistry did not work for tin.

Over coffee in the office, one of the staff mentioned how at University he had been told that cassiterite is insoluble in nitric and perchloric acid. Perhaps they were using the wrong analytical method—what they needed was a "Total Tin" technique. A phone call to the lab confirmed that cassiterite is essentially insoluble in the acids used, and the project geologist quickly asked if the samples could be rerun for "Total Tin" instead. The lab, following this clear instruction, reran the samples using a sodium peroxide fusion, which is as close to a total tin analysis as you can get by a chemical technique.

The results were far from encouraging (Figure 4.21). Soils over the whole area of the stock now carry raised levels of tin but values are not much higher than the acid extraction results—just more extensive. Worse again, the best indication of the known mineralization is a small peak at the site of the old dump. This was worrying since percent concentrations of tin occur at outcrop in the vein.

Another session over coffee yielded a further suggestion from the same geologist. He had read that cassiterite is quite inert (it forms placer deposits) and usually occurs in soils as free mineral grains. Surely the geochem survey could be improved by isolating and analyzing the heavy mineral fraction. This sounded most encouraging so the area was resampled, at no small expense, and the samples forwarded to the lab for processing. Once again it was the minus 80 mesh fraction that was separated and from which a heavy mineral concentrate was prepared and analyzed for total tin. The results showed yet another pattern (Figure 4.22). True, absolute values are higher, we are now dealing with hundreds and thousands of ppm, but there is still no sign of the known mineralization—just the dump.

All this was most discouraging and sentiment was growing that soil geochemistry did not work. The geologists remained convinced that the ground had potential. Undoubtedly there was tin on the property but the geochemistry had only served to confuse the issue by providing a series of conflicting distribution patterns. Since they could not see the mineralization in the data, they did not know what to believe. Their final suggestion was that there must be "an overburden problem, masking the mineralization".

Management was similarly distressed that considerable

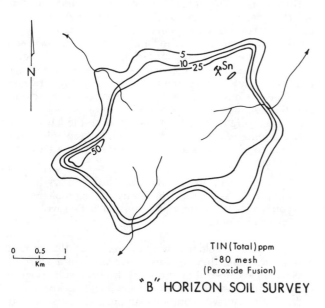

FIGURE 4.21—Tombstone Granite: Sn content (ppm) of minus 80 mesh B horizon soils using a peroxide fusion decomposition. This is effectively a total decomposition for Sn.

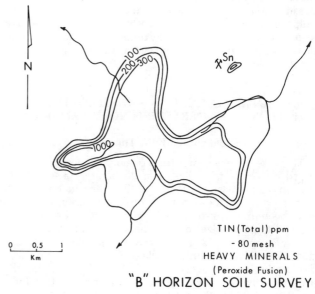

FIGURE 4.22—Tombstone Granite: Sn content (ppm) of minus 80 mesh heavy minerals in B horizon soils using a peroxide fusion decomposition.

time, money and effort had failed to find the promised target. But the area still looked good geologically, moreover there was an excellent land position. It was decided to spread the risk by joint venturing. Let me now introduce company number two—The Sourdough Land, Oil, Water and Mining Company. Affectionately known as SLOW MINING.

Slow Mining agreed that the area looked good but were not going to take the venture "sight unseen". They asked for the data and arranged a property visit during which their geologist and geochemist took rock and soil samples from the prospect area and at a few locations across the granite stock. These orientation samples were carefully studied by a variety of techniques and, on the basis of the results, Slow Mining negotiated a deal.

What had Slow Mining found that gave them encouragement? The property visit by their geologist and geochemist had confirmed the character of the mineralization and possible further mineral potential. Soils were found to be deep but residual over the entire area. Surface geochemistry such as B horizon soil sampling should reflect bedrock beneath.

The most important early information came with analysis of the rock samples taken from the prospect and granite outcrops. These were analyzed for contained tin by three techniques specified by the geochemist. Nitric-perchloric acid extractable and peroxide fusion (total) tin were determined to provide qualifying information on the work done by Smart–Ex. The rocks were also analyzed for tin following an ammonium iodide fusion, an extraction that is remarkably selective or specific to cassiterite.

Results (Table 4.17) show that, at the prospect, trace amounts of tin are acid soluble. This is almost certainly in the accessory mica in the vein. Almost all the tin occurs as cassiterite, more than 1% in the sample analyzed. The similarity between the ammonium iodide and peroxide fusion results confirm that essentially all the tin in the veins occurs in cassiterite.

Granite sample A was taken from an unmineralized outcrop near the prospect. It shows virtually all the tin in the rock to be acid soluble—presumable occurring largely in biotite, which makes up 4–10% of the rocks.

Granite sample B was taken close to the center of the stock within the broad total tin in heavy minerals pattern found by Smart–Ex. Here almost all the tin is extractable only by the total peroxide fusion. There is no evidence of acid soluble tin or significant cassiterite. The pattern indicates that the tin is in a resistate mineral: in this case accessory sphene.

Armed with these results the Slow Mining personnel were able to reinterpret the existing data from the Smart–Ex soil surveys. Thus the pattern for acid extractable tin in soils reflects tin in biotite and perhaps hornblende in one or two phases of the granite in the east and west. The geochemistry is here mapping one aspect of mineralogy and hence rock type. In the east this pattern maps the rock type hosting mineralization: could it be the case in the west?

Total tin relates to all tin in the bedrock and is roughly similar throughout the stock with a high in the west. In the bulk minus 80 mesh sample, we are looking at all available tin regardless of mineralogy. Total tin in the heavy mineral fraction reflects tin in all heavy minerals including casseriterite. Study of the pattern of tin values coupled with the work on the rock samples and microscopic examination of some heavy mineral concentrates confirmed that, over most of the anomalous area, this flat response is related to tin in accessory sphene disseminated through the granite. At least one further phase of intrusion is thus indicated by the data.

The increased tin values in the west looked interesting to the Slow Mining team. As well as being high values they are coincident with the area of acid soluble tin; the signature of the host to mineralization in the east. On completing the deal, Slow Mining had the minus 80 mesh heavy mineral samples collected by Smart–Ex rerun for tin by the cassiterite specific method (Figure 4.23). A tin target in the west is now clearly defined. Microscope examination of the concentrates revealed the presence of abundant topaz and fluorite—a geologically significant association.

Subsequent work confirmed the presence of mineralized

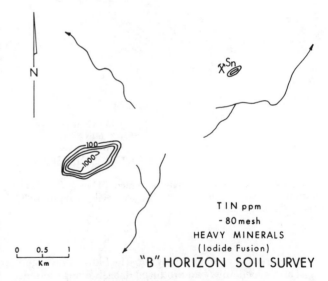

FIGURE 4.23—Tombstone Granite: Sn content (ppm) of minus 80 mesh heavy minerals from B horizon soils using an iodide fusion decomposition. This decomposition is specific for Sn as cassiterite.

TABLE 4.17—Tin in rock samples by three digestion procedures. All values in ppm unless otherwise indicated.

Digestion	Granite A	Granite B	Vein material
Nitric-perchloric acid	25	2	35
Iodide fusion	3	2	1.38%
Peroxide fusion	27	30	1.39%

greisen zones at the margins of a small leucocratic plug that itself carries sparse disseminated cassiterite. But there is still no sign of the known mineralization. This was, however, not a surprise to the Slow Mining geochemist. Comprehensive study of bulk soil samples collected around the prospect had shown that, in the B horizon, 98% of all the tin is in cassiterite grains coarser than 40 mesh. The Smart–Ex minus 80 mesh survey could never have found the tin mineralization. Luckily the oversize from the Smart–Ex heavy mineral survey was still available. For reasons quite unknown, they had been saved by the laboratory. They were secured by Slow Mining, reprocessed and analyzed by the cassiterite specific method. The tin data for the $-10+40$ mesh heavy mineral fractions are dramatically different (Figure 4.24). Not least the contours are in percent tin rather than parts per million. Moreover, the suspected but previously unknown high grade vein system is now revealed.

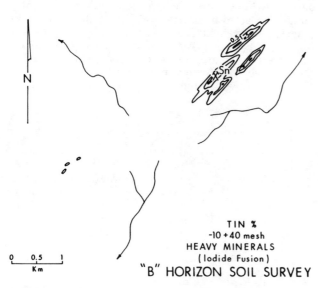

FIGURE 4.24—Tombstone Granite: Sn content (%) in $-10+40$ mesh fraction of heavy minerals from B horizon soils using an iodide fusion decomposition.

The work was most encouraging. The Slow Mining geologist and geochemist had produced data showing evidence of a multiple intrusive complex with significant tin mineralization of two types. The exploration effort was truly productive. Both venture partners were happy. This case history is a good and not at all unusual example of how many exploration programs fail because of unquestioned acceptance that standard procedures must always work. It is also a cautionary illustration of how a little knowledge can be very dangerous.

REFERENCES

Allcott, G.H. and Lakin, H.W. 1975. The homogeneity of six geochemical exploration reference samples. In: Elliott, I.L. and Fletcher, W.K. (editors), Geochemical Exploration 1974. Elsevier, Amsterdam, p. 659–681.

Brady, N.C. 1974. The Nature and Properties of Soils, 8th Edition. MacMillan, New York.

Chao, T.T. 1972. Selective dissolution of manganese oxides from soils and stream sediments with acidified hydroxylamine. Soil Science Society of America Proclamations v. 36, p. 764–768.

Clifton, H.E., Hunter, R.E., Swanson, F.J. and Phillips, R.L. 1969. Sample size and meaningful gold analysis. United States Geological Survey, Professional Paper 625–C, 17 pp.

DiLabio, R.N.W. 1985. Gold abundances vs. grain size in weathered and unweathered till. In: Current Research, Part A, Geological Survey of Canada, Paper 85–1A, p. 117–122.

Dolezal, J., Povondra, P. and Sulcek, Z. 1968. Decomposition Techniques in Inorganic Analysis, English Edition. Iliffe Books, London, 244 pp.

Filipek, L.H., Chao, T.T. and Theobald, P.K. 1982. Comparison of hot hydroxylamine hydrochloride and oxalic acid leaching of stream sediment and coated rock samples as anomaly enhancement techniques. Journal Geochemical Exploration, v. 17, p. 35–47.

Fletcher, W.K. 1981. Analytical Methods in Geochemical Prospecting. Elsevier, 255 pp.

Haffty, J., Riley, L.B. and Goss, W.D. 1977. A Manual on Fire Assaying and Determination of the Noble Metals in Geological Materials. United States Geological Survey, Bulletin 1445, 58 pp.

Hansuld, J.A., Mannard, G.W., Laikin, H.W., Canney, F.C., Salmon, M.L. and Weber, G.F. 1969. What is a geochemical analysis?—A panel discussion. Colorado School of Mines Quarterly, v. 64, p. 5–26.

Koksoy, M., Bradshaw, P.M.D. and Tooms, J.S. 1967. Notes on the determination of mercury in geologic samples. Institute Mining and Metallurgy Transactions, Section B, v. 76, p. 121–124.

Reedman, J.H. 1979. Techniques in Mineral Exploration. Elsevier Applied Science Publishers, England, 612 pp.

Shilts, W.W. 1984. Till geochemistry in Finland and Canada. Journal Geochemical Exploration, v. 21, p. 95–117.

Smith, R.E., Moeskops, P.G. and Nickel, E.H. 1979. Multi-element geochemistry at the Golden Grove Cu–Zn–Pb–Ag Deposit. In: Pathfinders and Multi-Element Geochemistry in Mineral Exploration. Extension Service, The University of Western Australia, p. 30–41.

Thanawut, S., Fletcher, W.K. and Dousset, P.E. Evaluation of geochemical methods in exploration for primary tin deposits: Batu Gajah—Tanjong Tualang area, Perak, Malaysia. Journal of Geochemical Exploration, in press.

Thompson, M. and Howarth, R.J. 1978. A new approach to the estimation of analytical precision. Journal Geochemical Exploration, v. 9, p. 23–30.

Thompson, M. and Walsh, J.N. 1983. A Handbook of Inductively Coupled Plasma Spectrometry. Blackie, Glasgow and London, 273 pp.

Ward, F.N., Nakagawa, H.M., Harms, T.F. and VanSickle, G.H. 1969. Atomic Absorption Methods of Analysis Useful in Geochemical Exploration. United States Geological Survey, Bulletin 1289, 45 pp.

Wargo, J.C. and Powers, H.A. 1978. Disseminated gold in Saddle Prospect, Lander County, Nevada. Journal Geochemical Exploration, v. 9, p. 236–241.

Chapter 5

STATISTICAL INTERPRETATION OF SOIL GEOCHEMICAL DATA

A. J. Sinclair

INTRODUCTION

The problem of data analysis is a combination of philosophy of approach and a clear understanding of the quality and nature of the data to be interpreted. To attempt to carry out blind evaluation of data by submitting them to any one of the ever increasing number of packaged software systems and expect a computer to do our thinking for us is patently wrong. Earth science data generally are collected with a particular goal in mind. The collection procedure ideally should incorporate an element of rigorous experimental design that provides efficiency in the accumulation of the data, representativeness of the data relative to the problem on hand and monitoring during data collection and analysis to provide a periodic measure of quality. Where such planning has gone into the design of a sampling program there generally is an advance appreciation of the interpretive methodologies that will be followed. Unfortunately, all data analysis situations are not planned in such a comprehensive manner. The other end of the spectrum is the situation in which a set of numeric data with relatively little accompanying information is provided with a request that an "interpretation" be forthcoming. Fortunately, most data interpretation exercises are between these two extremes, bounded on the one hand by high costs which control the element of design that can be incorporated in the undertaking, and, on the other hand, by the recognition that without some minimum amount of control the problem will be insoluble.

In applied geochemistry major programs that generate large amounts of quantitative and descriptive (categorical) data commonly are preceded by an *orientation survey* which utilizes a limited amount of effort and data to provide basic information useful in the efficient design of the broader program. In the case of soil data directed towards mineral exploration such a major program probably will be either of reconnaissance or detailed nature. In a general way, reconnaissance data are less regularly distributed whereas detailed data are more closely spaced and commonly arranged on a regular grid. The techniques used in interpreting these two extreme types of data may well be different. One obvious difference concerns the relative importance of a single sample. In the case of reconnaissance data each sample is apt to be independent of every other sample; thus, each sample is important in its own right. For detailed data such is not always the case and the clarity with which the interpretation of each individual sample is understood may not be so important.

Our data must be of appropriate quality for the purpose on hand, they must be representative (unbiased) and the measuring technique used to obtain the "numbers" must have adequate precision. Optimum data-collecting methods can be based on an orientation survey, as indicated previously.

This course is directed towards a rigorous evaluation of quantitative data of the types encountered in practical, mineral exploration-oriented soil surveys and emphasizes the use of statistical methods. It will not be possible to include all types of soil data in our discussion, just as it will be impossible to consider the pros and cons of all statistical methods. Emphasis will be placed on relatively simple techniques combined with a systematic progression leading to more complicated procedures (Figure 5.1). This approach is based on the philosophy that for most quantitative information, the obvious interpretations will emerge through rigorous examination of data by relatively simple methodologies. In some cases it will not be necessary or desirable to continue to complicated interpretive procedures. Whatever the ultimate purpose of a statistical study and no matter what complex methods be utilized, it is well to understand the individual variables, their histograms (probability density functions) and the extent to which simple methods can be used to understand the significance of these variables. Such an approach leads to a higher level of confidence and understanding of output from multivariate methods of analysis than might otherwise be the case.

Statistics is concerned initially with measures of (1) central tendency and (2) dispersion, as parameters that are useful in describing attributes of the probability density function of data. In practice, these parameters are calculated from a data set (a group of numbers) that represents a *sample* of a larger *population* under study. For example, 173 B horizon soil samples collected at regular grid intersections and analyzed for their Cu content represent a data set that is one sample of all possible B horizon soil samples that form the total population. A slight shifting of the grid would have provided a different suite of soil samples, that is, a different statistical sample. Note our dual usage of the word sample. Each soil sample represents an individual geochemical sample, but the total set of 173 soil samples represents a single statistical sample of size n = 173 items.

Ordinarily, we will find that our ability to extract useful

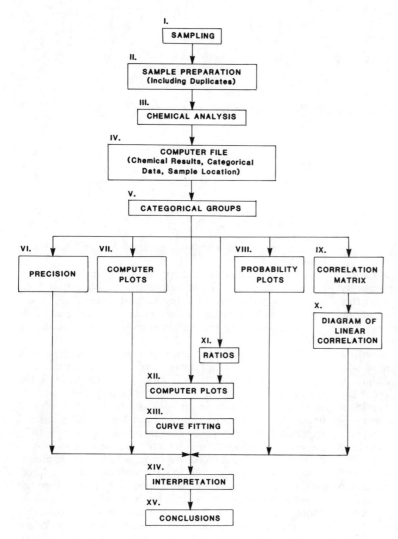

FIGURE 5.1—Flow chart illustrating a systematic approach to the evaluation of quantitative data from a soil geochemical survey. Only relatively simple techniques of data evaluation are represented. Note stage V which involves grouping data early in the evaluation process on the basis of appropriate categorized variables.

information from a data set will be enhanced if we are able to group data easily into various categories of geological significance. For example, we take great pains to distinguish and treat separately A, B and C horizon soil samples because to combine them is to obscure the effects of the genetic processes that lead to such fundamental attributes of soils. In a practical sense we unnecessarily complicate the interpretation procedure by combining data from such groups for which we have a priori evidence. Similarly, experience has shown that in many areas metal abundances in soils reflect parent material, so lithology of provenance becomes an important grouping criterion for data interpretation. It is possible to become very detailed in terms of attributes that might be used to group data so as to aid interpretation.

For soils such a list might include:

(1) soil classification
(2) soil horizon sampled
(3) possibility of contamination by man
(4) presence of colored chemical precipitates
(5) amount of organic matter present
(6) sand/clay ratio of sample
(7) pH at sampling site
(8) parent material from which sample was derived
(9) general physiographic description of environment around the sample site (e.g. steepness of slopes)
(10) vegetation type

The extent to which one records such information depends very much on the nature of a survey. Nevertheless, it is useful to bear in mind that early recognition and recording of factors that might be important in data interpretation (i.e., for data grouping) will facilitate the interpretation.

For many statistical techniques small data sets can be treated manually. For large data sets manual procedures are impractical and a computerized approach is necessary. Many statistical calculations cannot be done manually even for small data sets. Consequently, it is wise to organize data so they can be entered easily into a computer, whether such use is anticipated or not. Many forms have been developed for this purpose (e.g., Figures 3.7 and 3.8). Along with such forms a thoroughly unambiguous explanation of codes is required for the user, that is, a "user's manual" (e.g., Appendices I to III of Chapter 3).

BASIC STATISTICS

General Statement

The most fundamental aspects of statistics lie in measures of *central tendency* and *dispersion* of unbiased samples. These parameters commonly define the important characteristics of the probability density function of a data set and, if unbiased, estimate the comparable parameters of the parent population being studied. It is important to bear in mind that a sample never defines a population perfectly; that is, an error always exists, even though on rare occasions this error might coincidentally be zero. Our problem is, we rarely know the precise error, although, generally we can estimate the *average* error.

Central Tendency

Arithmetic mean

The arithmetic mean is perhaps the commonest measure of central tendency and is simply an average value of n items, determined by dividing the grand sum of all items (Σx_i) by the number of items (n).

$$\bar{x} = \frac{\sum_{i=1}^{n} x_i}{n}$$

We may need to know the mean Zn contents of the A and B horizon of soils in a target area to assist in deciding which of these has the greatest geochemical contrast, or, we might require mean Cu values for soils developed over two groups of metavolcanic rocks to decide if they are of equivalent or different chemical make-up.

Median

The median is a "central" value that divides an ordered set into two groups consisting of equal numbers of items. With an even number of items the median is intermediate between the lowest value of the upper group and the highest value of the lower group. For very small samples (n<5) the median is a more stable estimate of central tendency than is the mean.

Mode

The mode of a data set is the most abundant value. If data are grouped into classes, the mode is taken as the center of the most abundant class. If a mode is defined as a high relative to only the two contiguous values then a data set commonly will have more than one mode not all of which need be significant statistically. The recognition of multimodal populations is an important aspect of data evaluation.

Geometric mean

The geometric mean is the antilog of "the average of log transformed items in a data set".

Note that for many symmetric distributions of items about the mean value, the arithmetic mean, median and mode are commonly identical or nearly so. In very rare cases, two modes might be disposed symmetrically about a minimum that coincides with the mean and median.

Dispersion

Dispersion is a description of the spread of values, and is therefore an important attribute of various types of geochemical data.

Range

The range is defined by the two limiting values (high and low) of a data set and is normally represented by these two numbers rather than just their difference. The obvious reason for quoting extreme values is that otherwise one does not know the disposition of the range relative to the measure of central tendency. In general, the range is an unstable measure of dispersion because a single value can result in a drastic change in range. Similarly, because of a single outlying value the range may not describe satisfactorily the remaining data. With small data sets the range is a commonly quoted measure of dispersion; with very small data sets, the data themselves can be listed; with large data sets, other methods of describing dispersion (variance, standard deviation) are preferred.

Variance

An important standardized method of describing dispersion is the variance, which is the mean squared difference between individual items and the arithmetic mean of those items.

$$s^2 = \frac{\sum_{i=1}^{n} (x_i - \bar{x})^2}{n-1}$$

where s^2—estimate of population variance
$\bar{x}$—arithmetic mean of n items
x_i—represents successively the n items and
$n-1$—degrees of freedom in the denominator is necessary to provide an unbiased estimator of the population from the sample.

With large data sets, as might result from many soil geochemical surveys, it is apparent that estimation of s^2 is not critically dependent on division by $n-1$ as opposed to n. Note that the variance is a "squared" parameter so that no distinction is made between positive and negative differences used in its calculation.

Standard deviation

The standard deviation is the square root of the variance and is perhaps the most commonly quoted dispersion measure in statistical treatment of geochemical data. For routine discussion of dispersion, the standard deviation is generally more practical than the variance because the standard deviation is in the same units as the items themselves (e.g., ppm rather than ppm^2). Nevertheless, the variance is the more fundamental quantity.

Percentiles

Percentiles are values below which a stated proportion of a data set occurs. Various percentiles have been used as measures of dispersion for specific purposes, particularly in sedimentology in the analysis of sediment size distributions. Percentiles are becoming more widely used in geochemistry, especially in connection with probability graphs.

Some examples of percentiles are:

P_{10}, P_{90} values corresponding to 10 and 90 cumulative percent, respectively of the data.

P_{25}, P_{75} (or Q_{25}, Q_{75}) values corresponding to 25 and 75 cumulative percent of the data. Commonly referred to as quartiles.

P_{50} the median.

These percentiles have been used to define skewness and kurtosis of density functions, parameters that measure the departure of a distribution from a standard symmetric form.

Histograms

Histograms are a familiar method of displaying numerical information. Figure 5.2 shows three histograms illustrating common variations in form that occur in the case of populations encountered with mineral exploration data, metal abundances in soils in this case. Negatively skewed (Figure 5.2a), symmetric (Figure 5.2b) and positively skewed (Figure 5.2c) histograms are illustrated. Histograms are useful because they provide a simple visual display of (1) range of data, (2) modes, (3) general form of the probability density function, and (4) possible thresholds separating background and anomalous values. An additional advantage is that the preparatory grouping of data provides a relatively convenient form for manually calculating the mean and variance by the method of grouped data.

In constructing a histogram we must first choose an appropriate class interval between one-quarter and one-half the standard deviation of the data. If the class interval is too great the true form of the distribution is masked—if too small then too many gaps appear in the resulting histogram and the underlying form cannot be recognized. Secondly, the choice of where to start a class interval is not a serious matter as a rule but it seems sensible to standardize the procedure, by having two central classes disposed symmetrically with respect to the mean value.

It is useful to construct a histogram with the ordinate (frequency) as a percentage if comparison is to be made with other histograms with different sample sizes. It is good practice to include on a histogram or in the accompanying caption a listing of (1) title, (2) N—the sample size, (3) the class interval, and (4) the mean and standard deviation of the data.

Continuous Distributions

As the class interval of a histogram decreases for large samples, it becomes easier and easier to pass a smooth

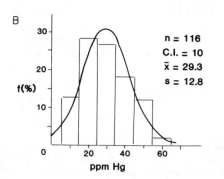
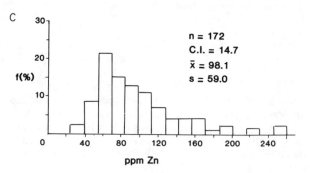

FIGURE 5.2—Examples of histograms of soil geochemical data sets: (a) Ba in B horizon soils, Daisy Creek strata-bound Cu prospect, western Montana (Stanley, 1984). (b) Hg in B horizon soils over the Daisy Creek stratiform copper prospect, western Montana. (c) Zn analyses of B horizon soils developed on a thin layer of ground moraine overlaying the Tchentlo porphyry Cu–Mo prospect in central British Columbia.

continuous curve through the tops of the classes. Consequently, it is possible to approximate many histograms of continuous or nearly continuous variables by a smooth mathematical curve known as a probability density function (pdf) (see Figure 5.2b). One might imagine that many such mathematical models would be required to take into account all potential pdf's of real data and while this might be true in theory it is fortunately not so in practice. A majority of variables in nature exhibit shapes of histograms that can be approximated by a relatively small number of mathematical models providing data are not truly multimodal. In fact, we will confine our attention here to two specific forms, the normal and lognormal distributions and will attempt to justify this position later in the present chapter.

The normal or Gaussian distribution was first put forward as a theory of error measurement. For example, we might wish to test the reproducibility of a particular chemical method of analyzing soils. From a large, single, well mixed standard, ten small subsamples might be taken, each of which is analyzed using the same method. The 10 values obtained will not necessarily be exactly the same due to random variations in analytical procedure. The spread of measured values about the mean follows what is known as the normal or Gaussian density function given by the following formula

$$y = \frac{1}{\sigma\sqrt{2\pi}} e^{-\frac{1}{2}(x-\mu)^2/\sigma^2}$$

where μ is the arithmetic mean, x is any measurement, and σ^2 is the variance of the population. The graphical expression is the familiar bell-shaped curve shown in Figure 5.2b.

Standard Normal Distribution

A normal pdf closely approximates many types of raw and log transformed geochemical data. Consequently, a knowledge of the pdf is fundamental to a formal treatment of much soil geochemical data.

All normal probability density functions are related by a simple transformation that reduces any such distribution to a standard form, the standard normal density function, in which a standardized value z is defined as

$$z_i = \frac{x_i - \bar{x}}{s}$$

In other words, each item is transformed into a new value which is "the number of standard deviations the original value is removed from the mean". Total area, A, under the symmetric curve is 1 and any line parallel to the ordinate axis divides this area into two proportions that sum to 1. The mean value, for example, divides A into two equal and symmetrically equivalent areas. The interval $\bar{x}-s$ and $\bar{x}+s$ encompasses 0.68 A, i.e., about 68% of the area with 34 % on each side of the mean. The interval $\bar{x}-2s$ to $\bar{x}+2s$ defines 0.95 A, or about 95% of the area (Figure 5.3). Tables have been prepared listing the proportions of area occurring from minus infinity to many positions on the standard normal curve, up to and including the mean (zero). Because the normal distribution is symmetric about the mean, proportions of area need be tabulated for only half the distribution. Note that with such a table it is possible to determine the proportion of area between asymmetrically distributed values such as $\bar{x}+1.6s$ and $\bar{x}-0.7s$ (i.e., the proportion of area up to $z = -0.7$ can be subtracted from the proportion of area up to $z = 1.6$ to give the area under the normal curve between these two abscissas, viz. $A_{1.6} - A_{0.7} = 0.94 - 0.24 = 0.70$.

Lognormal Distributions

In its simplest conceptual form the lognormal distribution is a normal distribution of the logarithms (to any base) of a set of data. Many earth science variables including minor elements in soils have histograms (pdf's) that are approximated closely by the lognormal law. For example, minor elements in geochemistry (e.g. Shaw, 1961); grades and tonnages of mineral deposits (e.g. Sinclair, 1974b); sediment size data (e.g. Harris, 1958), and so on. In addition, some variables dealt with routinely by earth scientists have normal *pdf's*, but the very nature of the variables incorporates a log transformation: pH measurements and sediment size data (in phi units) are everyday examples.

Of course, not all chemical variables in soils are lognormally distributed. Although many trace elements have positively skewed distributions that can be approximated by a lognormal distribution, others have histograms with different forms. It is not abnormal for chemical constituents of soils with mean values in the range 1 to 10 percent to approximate a normal distribution; analyses of materials that approach a pure mineralogical composition may have a negatively skewed distribution as with Al contents of lateritic soils.

Real data depart most from continuous empirical models at the tails of the fitted distributions. Exactly where a given model no longer applies to real data is difficult to determine but one can be fairly certain that a lognormal model, for example, cannot be applied with assurance beyond the range of data on which the model is based.

Perhaps the most serious difficulty in reconciling a lognormal model with real data is encountered with polymodal distributions. Where component populations do not overlap appreciably each can be examined individually for lognormality. The frequency of occurrence of such polymodal lognormal distributions indicates that it is logical to expect

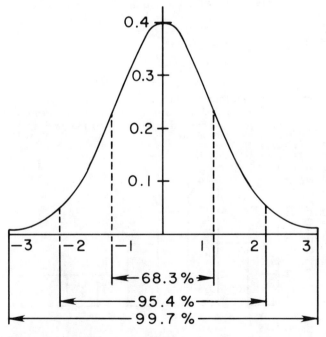

FIGURE 5.3—Standard normal curve. Percentages of areas under various segments of the curve are shown.

overlapping populations to also approximate lognormal models. Certainly, this latter approach has proved practical as a working hypothesis (e.g. Montgomery et al., 1975; Saager and Sinclair, 1974).

In some cases it is necessary to estimate arithmetic parameters from a distribution whose parameters are known in logarithmic units, or vice versa. The following equations can be used.

From natural log parameters to arithmetic parameters

$$\bar{x} = be^{s_n^2/2}$$

and

$$s^2 = \bar{x}^2\left(e^{s_n^2} - 1\right)$$

where x̄—arithmetic mean
s—arithmetic standard deviation
b—geometric mean (antilog of $\bar{x}_n$, the mean of the natural logarithms)
s_n—standard deviation of natural log transformed values (in natural log units)

From logarithmic (base 10) parameters to arithmetic parameters

$$\bar{x} = b10^{(1.1513 s_t^2)}$$

and

$$s^2 = \bar{x}^2\left[10^{(2.3026 s_t^2)} - 1\right]$$

where b—geometric mean (antilog of $\bar{x}_t$, the mean value of $\log_{10}$ values)
s_t—standard deviation of $\log_{10}$ transformed values (in $\log_{10}$ units)

The antilog of a mean value of log transformed data is constant regardless of the base of the logarithms. Thus, in the foregoing equation b is identical whether estimated as the antilog of $\bar{x}_n$ or $\bar{x}_t$.

A natural logarithm is about 2.3026 times larger than the $\log_{10}$ transform for the same number. Thus, ln(25) = 3.2189 and $\log_{10}(25)$ = 1.3979. These calculations are illustrated by a simple example. Consider the 15 numbers tabulated in Table 5.1 and illustrated as a histogram in Figure 5.4. Means and standard deviations for raw data, log transformed data and natural log transformed data have been calculated separately and are listed. Now let us use the two sets of log-based parameters to estimate the arithmetic parameters. Substitution in the equations presented earlier gives the estimates shown in Table 5.1. There is a discrepancy of about 2% of the true mean value and 23% in the standard deviation. These apparent discrepancies arise from the fact that the small data set selected is discrete and is not exactly lognormal in form but is simply an arbitrarily chosen positively skewed data set that is crudely approximated by a lognormal distribution.

Fitting a normal curve to a histogram

Once a histogram is obtained we can test qualitatively for normality in two ways: (1) plotting cumulative data on probability paper (considered in a later section), and (2) con-

TABLE 5.1—Hypothetical data illustrating estimation of raw data parameters from log parameters.

Data items		
1	3	5
2	3	5
2	4	6
3	4	7
3	4	8

	Parameters	
Arithmetic	Natural logs	log10
x̄ = 4.0000	$\bar{x}_n$ = 1.2650	$\bar{x}_t$ = 0.5494
s = 1.9272	s_n = 0.5379	s_t = 0.2336
	b = 3.5431	b = 3.5432

b is the geometric mean

Arithmetic parameters calculated from log parameters		
	Derived from*	
	Natural logs**	log10***
x̄	4.0946	4.0947
s	2.3718	2.3718

*Assumes pdf is lognormal
**$\bar{x} = b\, e^{s_n^2/2}$
$s^2 = \bar{x}^2(e^{s_n^2} - 1)$
***$\bar{x} = b\,(10)^{1.1513\, s_t^2}$
$s^2 = \bar{x}^2\,[(10)^{2.3026\, s_t^2} - 1]$

structing a normal curve with the same parameters (mean and variance) as the data.

The equation for fitting a normal curve to a set of data is:

$$y_i = \frac{n \cdot i}{s}\left[\frac{1}{\sqrt{2\pi}}e^{-\frac{1}{2}[(x_i - \bar{x})/s]^2}\right]$$

where n is the number of data points or values, i is the class interval of the histogram, and the exponential term and constant is obtained from the standard normal tables. It is

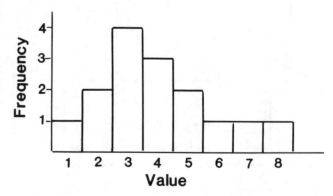

FIGURE 5.4—Histogram of 15 values listed in Table 5.1.

TABLE 5.2—Points on normal curve fitted to histogram for Hg in B horizon soils, Daisy Creek strata-bound Cu prospect, Montana (Stanley, 1984).

Abscissa		Ordinate	
Standard normal	Absolute	Standard normal	Absolute*
$\bar{x} + 2.55$	61.23	0.0175	1.37
$\bar{x} + 2.05$	54.85	0.054	4.23
$\bar{x} + 1.55$	48.46	0.130	10.18
$\bar{x} + 1.05$	42.08	0.242	18.95
$\bar{x} + 0.55$	35.61	0.352	27.57
$\bar{x}$	29.31	0.399	31.25
$\bar{x} - 0.55$	22.93	0.352	27.57
$\bar{x} - 1.05$	16.54	0.242	18.95
$\bar{x} - 1.55$	10.16	0.130	10.18
$\bar{x} - 2.05$	3.78	0.054	4.23
$\bar{x} - 2.55$	−2.40	0.0175	1.37

*$Y_{Abs} = Y_{Sta}(n \cdot i)/s$ where n = 100, i = 10, s = 12.77

most practical to calculate y_i values for x_i values that are separated by 0.5s, plot the y_i values and join them by a smooth curve, recognizing that inflection points occur at $\bar{x} \pm s$. An example of curve fitting is illustrated in Figure 5.2b and Table 5.2.

Cumulative distributions

Data prepared for a standard histogram can also be presented as a cumulative histogram in which, to the frequency within any class is added the total frequencies of all preceding classes. Frequencies can be cumulated from either the high or low end of the range of values. This method of representation is common in the field of sedimentology for percentage weights of sediment size fractions, where frequencies are cumulated from coarse-grained to fine-grained fractions. A geochemical example is shown in Figure 5.5 cumulated from low to high values with a superimposed curve illustrating cumulation of the same data from high to low values.

This method of cumulative frequency representation of data is in the required form for plotting on probability graph paper. The concept of a cumulative histogram is straightforward and is fundamental to an understanding of probability plots.

Confidence limits

Area, as a proportion, under a specific part of a normal curve is equivalent to the probability that a randomly drawn item from the population will lie within the range for the proportion of area in question.

Of course, parameters determined from a sample of a normal population are only estimates of the true parameters of the population. It is common procedure to place confidence limits on the estimated parameters, particularly the mean value. Where normal distribution parameters are estimated by a large number of items (n>120) the dispersion of mean values (s_e) for sample size n is itself normal and is given by

$$s_e = \frac{s}{\sqrt{n}}$$

where s is the sample standard deviation, n is the number of items, and s_e is the standard error (dispersion) of the mean, i.e. the sample mean has a normal distribution with dispersion (standard deviation) equal to s_e. Consequently, s_e can be used to put confidence limits on the mean. For example, there is 95% chance that the true mean lies between $\bar{x} - 2s_e$ and $\bar{x} + 2s_e$. Conversely, if we accept these 95 percent confidence limits as containing the true mean, we will be wrong 5% of the time.

For small sample size (n<120) the sample mean has a t-distribution and confidence limits are as follows (one standard error or 68% confidence limits):

$$(\bar{x} - t_\alpha s/\sqrt{n}) < \mu < (\bar{x} + t_\alpha s/\sqrt{n})$$

where $\bar{x}$ is the sample mean, μ is the population mean, $s/\sqrt{n}$ is the standard error, and the t_α value is obtained from t tables for given degrees of freedom (d.f. = n−1) and an acceptable α. Alpha is the type 1 error, that is, the probability that we will be wrong. In the case of 95% confidence limits we will be wrong 5% of the time so α is 0.05.

F and t tests

Of the many forms of hypothesis testing, our concern here is with the common problem of comparing two samples to test rigorously whether or not they could be drawn from the same population. If it is highly unlikely the samples represent the same parent population, we might *assume* they represent fundamentally different populations. Such tests are run at a selected error level, say, α = 0.05. In

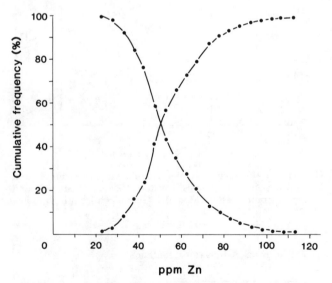

FIGURE 5.5—Cumulative curves for B horizon soil Zn analyses, Daisy Creek strata-bound copper prospect, western Montana. Curves are shown cumulated from low to high values and from high to low values.

general, these tests require normality of the variables being compared and begin with the null hypothesis, that is, the assumption that the two samples do indeed represent the same population. If data are not distributed normally an appropriate transformation may be necessary before F and t tests are done.

As an example of F and t tests consider the data of Table 5.3 from a soil geochemical survey in the Ashnola area of southern British Columbia (Montgomery et al, 1975). The F value of 1.14 is calculated from the data with the implicit assumption that data are distributed normally. For $\alpha = 0.05$ critical values of F (202,202) are 0.7 and 1.2. Because our calculated value lies within the critical range we conclude that, within limitations of our sampling program, the two populations have variances that are indistinguishable. In order to test whether or not the mean values are identical we estimate a pooled variance as follows:

$$s_p^2 = \frac{(n_1 - 1)s_1^2 + (n_2 - 1)s_2^2}{(n_1 + n_2 - 2)}$$

For Ashnola zinc data, s_p, the pooled variance is 14657 to give a pooled standard deviation of 121.1. The t test can now be conducted by calculating

$$t = \frac{\bar{x}_1 - \bar{x}_2}{s_p\sqrt{1/n_1 + 1/n_2}}$$

Calculated t for Ashnola zinc data is 0.59. For $\alpha = 0.05$ the critical value of t_α is 1.12 for 404 degrees of freedom. Our calculated value is substantially less than the critical value and we are led to conclude that the two means are indistinguishable. More generally, zinc data do not show any significant differences in mean value or dispersion in the A and B soil horizons of the survey area.

A paired t-test is substantially more rigorous than a t-test and should be used wherever possible. For a paired t-test we examine the distribution of real *differences* between paired values and test the mean difference between pairs to determine if this mean difference itself differs significantly from zero. This test is conceptualized most easily by determining if "zero" is inside or outside the 95 percent confidence limits of the mean difference.

Probability graphs

Probability paper is a useful practical tool in the analysis of soil geochemical data because of the common normal or lognormal character of such data. One ordinate of the graph paper is either equal interval (arithmetic) or logarithmic as required; the other, the probability scale, is arranged such that a cumulative normal (or lognormal) distribution will plot as a straight line. This type of graph paper is very sensitive to departure from normality and therefore to the recognition of combinations of multiple populations, a particularly useful attribute in dealing with soil and other types of geochemical data (e.g. Sinclair, 1974a, 1976; Parslow, 1974; Bolviken, 1971; and Lepeltier, 1969). An important consideration is the ease with which the method can be used in the field.

Data grouped for purposes of constructing a histogram can be cumulated (from high to low values or vice versa) as for a cumulative histogram, and plotted directly on probability graph paper. Here values are cumulated from high to low (cf. Lepeltier, 1969). Two examples are shown in Figure 5.6 where the straight line indicates a single lognormal population, and the curved line a combination of two lognormal populations.

A straight line can be fitted easily by eye to appropriate data and provides direct estimates of mean value and standard deviation of the logarithms of the data

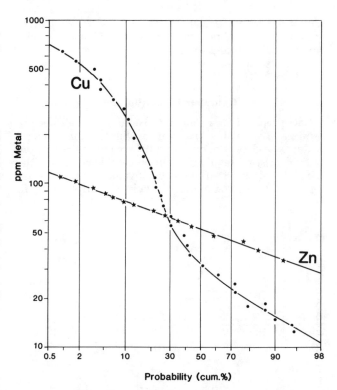

FIGURE 5.6—Probability graphs (cumulative curves) for Zn and Cu in B horizon soil samples over the Daisy Creek strata-bound copper prospect, western Montana. Cumulated from high to low values.

TABLE 5.3—Comparison of Zn analyses from A and B horizons, Ashnola area, southern British Columbia.

Variable	n	x̃	s	s²	F
Zn in A horizon	203	105	125	15625	1.14
Zn in B horizon	203	100	117	13689	

$$\bar{x} = P_{50}$$
$$s = (P_{16} - P_{84})/2$$
$$= (P_{2.5} - P_{97.5})/4$$

where P_n is the log value at the nth percentile. Graphical estimates of $\bar{x}$ and s for $\log_{10}$ data defining the straight line (Zn data) in Figure 5.6 provide arithmetic estimates of 54.9 and 16.8, respectively using formulas of the section "Lognormal Distributions". These compare with values calculated by the method of moments of 54.01 and 17.45, respectively.

Combinations of different proportions of two lognormal populations produce graphs similar in form to the z-shaped curve of Figure 5.6. Similar curves can be constructed graphically for any two populations A and B (each of which is a straight line) drawn on probability paper by repeated application of the formula

$$P_m = f_A P_A + f_B P_B$$

at various ordinate levels, m. P_m is a point on the combined plot, P_A and P_B are the cumulative percentages of the A and B populations respectively, and f_A and f_B (where $f_A = 1 - f_B$) are the corresponding proportions of the two populations. Note the ease with which any mixture of populations can be constructed by using different starting populations and varying their ratios in the mixture (see Figure 5.7). Examination of many plots shows that an inflection point occurs in the cumulative curve for a mixture at a cumulative percentage that coincides with the amounts of the two populations present.

In general, we work in a reverse sense. For soil Cu data of Figure 5.8 an inflection point is apparent at the 15 percentile indicating 15 percent of an upper lognormal population A, and 85 percent of a lower lognormal population B. Thus, should we want to partition the mixture into its individual components A and B by application of the above formula, we are faced with two unknowns, P_A and P_B, one of which must be known in order to calculate the other. Examination of many hypothetical mixtures shows that this form of curve results where two lognormal populations overlap partially. In other words, the high extremity of the curve represents variable amounts of A and negligible amounts of B. Furthermore, at the lower extremity of the mixture we have accumulated essentially 100 percent A plus variable amounts of B. Consequently, in the two extremities of the probability curve our equation has been reduced to a single unknown. Thus, at the upper end we can calculate points on the upper part of the A population and on the lower end, we determine points on the lower end of the B population. If the "ends" of these individual populations are defined clearly, they can be extrapolated into parts of the graph where they cannot be estimated directly. Extrapolation is commonly straightforward if the populations are

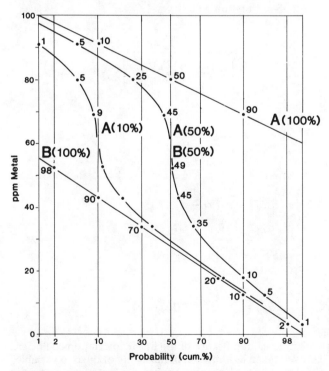

FIGURE 5.7—Probability graph of two ideal lognormal populations (A and B) combined in the proportions A:B = 50:50 and A:B = 10:90. Numbers on curves represent cumulative percent from high values to low for the upper population and from low values to high for the lower population.

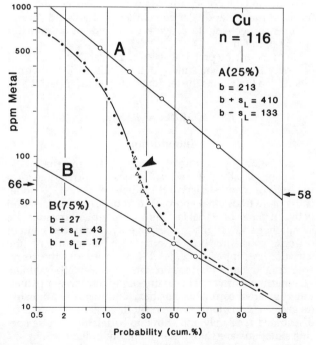

FIGURE 5.8—Probability graph of Cu in soils (from Figure 5.6) partitioned into two ideal components, A and B. An arrowhead shows the interpreted position of an inflection point in the curve. Black dots are original data; open circles are construction points determined by the partitioning procedure; open triangles are check points determined by combining ideal populations A and B in their zone of overlap to compare with real data. Thresholds of 58 and 66 ppm are determined at the lower 2.5 percentile of A and the upper 2.5 percentile of B.

lognormal in which case they are represented by *straight lines* on probability paper.

Open circles in Figure 5.8 are partitioning values obtained by applications of the partitioning formula to the individual data points used to define the "mixing" curve. A straight line has been fitted by eye to each set of partitioning points and then projected across the entire graph to provide estimates of the A and B populations. These ideal populations were then recombined to compare the ideal mixing curve with the raw data curve. Check points shown as open triangles are in extremely close agreement with the raw data curve and provide an internal check on the consistency of the partitioning model.

Partitioning of an apparent bimodal population is useful in providing a precise model against which to test real data. To do this most effectively it is convenient to pick practical thresholds (Sinclair, 1976, 1974a) that separate a bimodal distribution into three categories, viz. (1) essentially pure A, (2) essentially pure B, and (3) an intermediate group containing known proportions of A and B. In cases of no effective overlap of A and B populations the intermediate group (3 above) is not present. If A is anomalous and B is background such thresholds serve to assign priorities for followup investigation. More generally, the technique permits recognition of two ranges that are "pure" A and "pure" B, respectively. Color-coding of such data may provide insight into the significance of each population.

For the soil data of Figure 5.8 thresholds are chosen arbitrarily at the 2.5 and 97.5 percentiles of the B and A populations respectively to provide thresholds at 66 ppm Cu and 58 ppm Cu.

This account of practical uses of probability plots is of necessity limited. For a more detailed discussion of applications and limitations the reader is referred to Sinclair (1976).

CORRELATION

Introduction

Correlation is a measure of *similarity* between paired data. Two conceptually different categories of correlation form the basis of many statistical treatments. The first, R-mode, is the more traditional approach that deals with correlation between pairs of variables. As an example, consider the sympathetic variations in specific gravity and iron content of rocks consisting only of quartz and hematite; or, the variation in copper contents of soils relative to the corresponding organic matter contents. We want to examine whether the two variables in question increase sympathetically (positive correlation) or have an inverse relationship (negative correlation). Conversely, the variables might be distributed randomly and show no correlation. These specific situations are shown schematically in Figure 5.9.

The simple linear correlation coefficient (r) lies between -1 and $+1$, where an absolute value of 1 means perfect correlation and a value of zero means no correlation. A quantitative estimate of the simple linear correlation coefficient is

$$r = \frac{\text{cov}_{xy}}{s_x \cdot s_y}$$

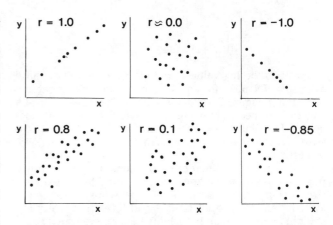

FIGURE 5.9—Schematic scatter plots to provide a conceptual interpretation of values of the simple linear correlation coefficient.

where s_x and s_y are the standard deviations of variables x and y respectively, and cov_{xy}, the covariance of the two variables can be determined as follows

$$\text{cov}_{xy} = \frac{1}{n}\sum_{i=1}^{n}(x_i - \bar{x})(y_i - \bar{y})$$

The formula for r can also be expressed in terms of "sums of squares" notation as follows

$$r = \frac{\sum xy - \frac{\sum x \sum y}{n}}{\left[\sum x^2 - \frac{(\sum x)^2}{n}\right]^{\frac{1}{2}}\left[\sum y^2 - \frac{(\sum y)^2}{n}\right]^{\frac{1}{2}}}$$

or

$$r = \frac{SS_{xy}}{\sqrt{SS_x \cdot SS_y}}$$

where

$$SS_x = \sum_{i=1}^{n}(x_i - \bar{x})^2$$

$$SS_y = \sum_{i=1}^{n}(y_i - \bar{y})^2$$

$$SS_{xy} = \sum_{i=1}^{n}(y_i - \bar{y})(x_i - \bar{x})$$

Testing of significance of the linear correlation coefficient is based on an assumption of normality of the two variables involved. Critical (absolute) values for r are reproduced for the 0.05 and 0.01 levels of significance in Table 5.4. These tables are entered using d.f. = $n-2$, where n is the number of pairs of items. For normally distributed variables a calculated simple correlation coefficient also can be tested for

TABLE 5.4—Critical values of the simple linear correlation coefficient for alphas of 0.05 and 0.01.

d.f.*	Critical r value		d.f.	Critical r value	
	$\alpha = 0.05$	$\alpha = 0.01$		$\alpha = 0.05$	$\alpha = 0.01$
1	0.997	1.000	20	0.423	0.537
2	0.950	0.990	22	0.404	0.515
3	0.878	0.959	24	0.388	0.496
4	0.811	0.917	26	0.374	0.478
5	0.755	0.875	28	0.361	0.463
6	0.707	0.834	30	0.349	0.449
7	0.666	0.798	32	0.339	0.436
8	0.632	0.735	34	0.329	0.424
9	0.602	0.735	36	0.320	0.413
10	0.576	0.708	38	0.312	0.403
11	0.553	0.684	40	0.304	0.393
12	0.533	0.661	50	0.273	0.354
13	0.514	0.641	60	0.250	0.325
14	0.497	0.623	70	0.232	0.302
15	0.482	0.605	80	0.217	0.283
16	0.468	0.590	100	0.195	0.254
17	0.455	0.575	125	0.174	0.228
18	0.444	0.561	150	0.159	0.208
19	0.433	0.549	200	0.138	0.181

*d.f. = degrees of freedom = number of pairs − 2.

"difference from zero" using graphs such as those provided by Krumbein and Graybill (1965). The null hypothesis is that our calculated r is determined for two independent and normally distributed random variables, and therefore is zero. A data transformation may be necessary to meet this criterion. Regardless, it is essential to view scatter diagrams where the nature of the correlation is a matter of great importance. Some potential problems are illustrated in Figure 5.10.

As the name implies, the linear correlation coefficient is a measure of the extent to which paired data plot as a straight line in a scatter diagram. It may be that some other kind of trend (e.g. quadratic) is present. Additional problems can arise from outlier values, two extremes of which are illustrated in Figure 5.10. In one case, essentially random data plus a single outlier lead to a high calculated correlation coefficient. In another case highly correlated data plus an outlier lead to a low correlation coefficient.

The second category of similarity measure common in earth science applications is Q-mode, which deals with correlation between samples (or stations). As an example, consider Co and Ni values determined at three sites A, B, and C illustrated in Figure 5.11. Each sample site can be considered a vector rather than a point. Samples B and C are very similar in their Co/Ni ratios as shown by the closeness of approach of their two vectors (i.e., the angle between the two vectors is small). On the other hand, sample A is appreciably different from both B and C.

A measure of correlation (relative similarity) between pairs of samples is given by the cosine of the angle (theta) between each pair of vectors, to produce a range of absolute cos (theta) values from 0 to 1, as in the case of the simple linear correlation coefficient. It is easy to see that this concept of Q-mode correlation can be extended to vectors in a 3-dimensional coordinate system (i.e. where we have measured

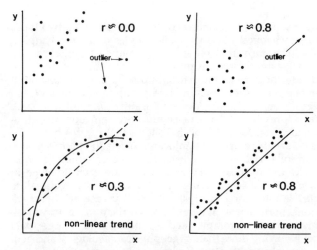

FIGURE 5.10—Schematic representation of potential problems with blind acceptance of simple linear correlation coefficients.

three variables at each site). Although more difficult to visualize, the same procedure can be applied to n dimensions, and it is here that the great importance of Q-mode correlations lies.

Analysis of a matrix of correlation coefficients

Correlation coefficients for paired variables are commonly presented as a two-dimensional array or matrix, which, if large, can create confusion rather than aiding interpretation of data. A useful procedure for analyzing such a matrix is to have the individual variables arranged in a meaningful manner. A simple example is illustrated in Table 5.5, where

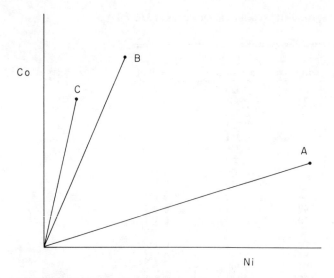

FIGURE 5.11—Representation of samples as vectors in n-dimentional space (2-D in this case) and use of the cosine of the angle (theta) between pairs of vectors as a measure of similarity (Q-mode) that is comparable in range to the simple linear correlation coefficient (R-mode). B and C are very similar (high cos theta value) whereas A and C are very different (low cos theta value).

the purpose is to examine correlations involving two types of geochemical data, viz. soil samples and rock samples. Variables measured on soils are grouped together, as are variables measured on rocks. Lines can then be drawn to divide the correlation matrix into subgroups representing *intra* group correlations and *inter* group correlations. Within each group those r values significantly different from zero (at a preselected level of significance) are underlined. A rapid visual scan of Table 5.5 shows that for this particular data set there are no significant correlations *between* rock and soil variables. The only significant correlations are intra rock, and of these, copper and zinc are each involved in two. A sensible grouping of variables used to generate a correlation matrix can result in relative ease of interpretation, particularly, where large correlation matrices are involved. The example cited is a part of much larger data set involving more than 25 geophysical, geological and geochemical variables (see Godwin and Sinclair, 1979).

There are two useful graphical methods of displaying correlation coefficients. The first is a correlation diagram where each variable is represented by a circle (Figure 5.12). For significant correlations the corresponding circles are placed close together and joined with a line on which the correlation coefficient is written. It is useful to construct a correlation diagram by beginning with the highest r value and progressing to the lower (but statistically significant) values. In this way, separate clusters of intra-correlated variables are seen to evolve. The correlation matrix can be reorganized to emphasize the various intra correlated groups if desired. For large numbers of variables correlation diagrams can become complicated; separation of variables into subgroups and construction of separate correlation diagrams for each subgroup may be desirable. A correlation diagram is shown in Figure 5.12 for the correlation matrix of Table 5.5.

The second method of representing correlation coefficients graphically is the dendrogram or the slightly different dendrograph. These diagrams are constructed most easily by computer. They are rarely constructed by hand except in the case of very small correlation matrices. Nevertheless, a manual example of a dendrogram is most useful to appreciate the diagrams and their limitations.

Consider the matrix of Table 5.5, which incorporates both positive and negative correlation coefficients. We might adopt the following procedure in constructing a dendrogram:

(1) Isolate the highest positive r value, here 0.605 (Pb vs. Zn)
(2) Calculate a new average r value, $(r_{1k} + r_{2k})/2$, between variables x_1 and x_2 on the one hand and all other variables x_k on the other
(3) Take the highest of the remaining original r and new average r values, in this case 0.595, to form a separate group of intra correlated variables (Cu vs. Mo)
(4) Now we determine the average intergroup correlation, i.e., $(0.180 + 0.595 + 0.120 + 0.201)/4 = 1.096/4 = 0.274$.

In this procedure we have ignored the effect of negative values. This can be taken into account by changing all correlation coefficients by an arc cos transform, i.e. "the angle whose cosine is", to produce a matrix of positive trans-

TABLE 5.5—Correlation matrix of selected rock and soil elements, Casino area, Yukon, Canada (after Godwin and Sinclair, 1979).

Variable		Soil			Rock			
		W	Ag	Au	Cu	Mo	Pb	Zn
Soil	W	1.000						
	Ag	0.035	1.000					
	Au	−0.071	0.121	1.000				
Rock	Cu	0.049	−0.003	0.027	1.000			
	Mo	0.096	0.166	0.054	*0.417*	1.000		
	Pb	−0.023	0.020	0.013	0.180	0.120	1.000	
	Zn	−0.115	−0.059	−0.023	*0.595*	0.201	*0.605*	1.000

n = 125
Values greater than 0.228 are significant at the 1% level.

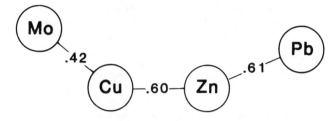

FIGURE 5.12—A simple correlation diagram for the matrix of simple linear correlation coefficients of Table 5.5. Only those simple linear correlation coefficients significant at the 0.05 level are reproduced between circles representing the variables involved.

formed values all in the range 0 to 180 (see McCammon and Wenninger, 1970, for details).

The resulting dendrogram (Figure 5.13) shows two groups, viz. Pb–Zn and Cu–Mo, which are weakly correlated. In fact, the diagram obscures the fact that both Cu and Pb are almost equally strongly correlated with copper. This problem is overcome somewhat in the dendrograph where variables are *not* equally positioned along the coordinate but are spaced somewhat in proportion to the magnitude of correlation with other variables. Thus, the dendrograph emphasizes groupings of elements to a greater extent than does a dendrogram. Both suffer from an impracticality of manual construction if a large number of variables are involved, and both contain hidden problems in dealing with negative correlations.

"Correlation" of populations

A common problem in geochemistry is the integration of information from several surveys, for example, the integration of two sets of soil survey data for contiguous areas. The two sets might be interdispersed, or might represent two separate areas. A problem arises in merging these data if the samples have been analyzed by different procedures that extract different proportions of a given variable. One simple solution that may be satisfactory is to look at probability plots of the two sets independently, determine thresholds for each of the two data sets, and construct a contour that encompasses all those values above the appropriate thresholds. Such a procedure implies a correlation between populations in the two sets of data, but does not require that the correlation be estimated quantitatively. This procedure was used by Montgomery et al. (1975) to combine data from two surveys over a porphyry copper system. An inconsistency in absolute metal values appeared between the two sets of data: but each data set consisted of 3 populations that appeared comparable in the two cases. Hence, thresholds were *equated* for contouring purposes even though they were numerically different.

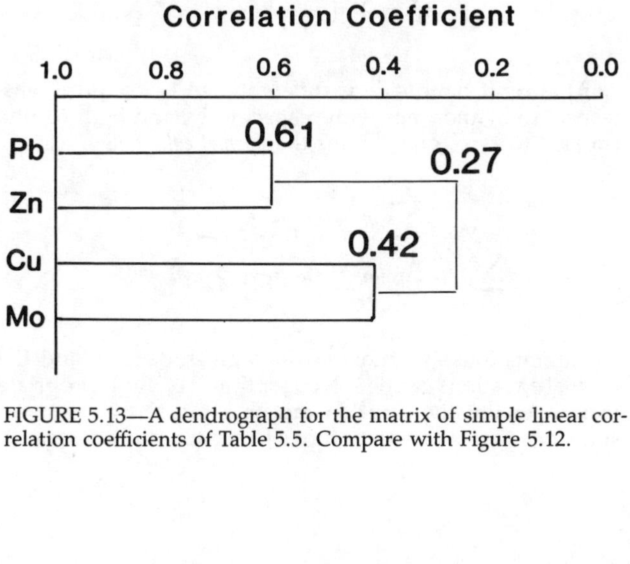

Correlations among percentage data

Many variables are expressed in percentage form, and totals necessarily are fixed (100%). Such situations lead to difficulties in the interpretation of correlation coefficients. Closed number systems are those that sum to a fixed amount; open systems are not so constrained. It is apparent that as the percentage of one member of a two-component system decreases, the amount of the other member must increase. This results in a perfect dependence of one variable on the other. The restraints become less apparent as the number of components increases.

The restraints in three-component closed systems are:

(1) r is completely controlled by the variances of the closed variables.
(2) two r's must be negative.

These restraints exist whether they were present in the original open data or not. The closed array correlation coefficients, r_{ij}, can be forecasted by

$$r_{ij} = \frac{1}{2} \frac{s_k^2 - (s_i^2 + s_j^2)}{s_i s_j}$$

where subscripts i, j, and k refer to the three components and s's refer to standard deviations of *open* data.

Autocorrelation

Autocorrelation concerns the correlation of a variable with itself, that is, correlation between paired items removed from each other either in time or in space. In dealing with autocorrelation in space we speak of a *regionalized variable* as opposed to a random variable which shows no spatial correlation. The theory of regionalized variables, so-called geostatistics of the French School, was developed principally as a theory fundamental to ore reserve estimation problems (Matheron, 1963) but has much wider application than implied by this narrow view of geostatistics.

Two standard approaches are used to study autocorrelation of geochemical data—the *correlogram* (or covariogram) and the *semivariogram*. Agterberg (1965) was one of the first to propose the use of serial (spatial) correlation in studying geochemical data. Spatial correlation can be thought of as a series of simple linear correlation coefficients, one for each sample spacing. Thus, in a regularly spaced sample array with sample spacing (lag) of one unit, all pairs separated by one unit are considered x and y, that is, if the first point is considered x, the second sample is y. When the second

FIGURE 5.13—A dendrograph for the matrix of simple linear correlation coefficients of Table 5.5. Compare with Figure 5.12.

sample is x, the third sample becomes y; then the third sample is x and the fourth is y; and so on. In this way spatially distributed samples are used to obtain paired data for which a correlation coefficient can be calculated. Of course the same procedure can be used for many different sample spacings and for sample lines oriented in different directions. In a regular grid, for example, pairs can be obtained independently in either of the principal grid directions and along the principal diagonals of a grid to permit study of variations in spatial correlation as a function of direction. If differences exist as a function of direction we say the structure of the variable is *anisotropic*; if no differences exist the variable is *isotropic*.

The correlogram is simply a plot of correlation coefficients as a function of sample spacing (lag). With many kinds of data nearby samples might be very similar whereas widely spaced samples are likely to be much less similar. A simple geochemical model that corresponds to such a view is that high values occur in anomalous zones whereas low values occur in background zones. With small sample spacings most adjacent sample pairs are either both in background or both in an anomalous area, consequently short lag samples are highly correlated. Where lag is greater than the dimensions of an anomalous zone, two samples cannot occur within an anomalous zone. Consequently, there is a greater proportion of large disparities between sample pairs, i.e. they are less correlated or perhaps not correlated at all.

A standard procedure, therefore, is to test each value of r as it is determined for its corresponding lag, to see whether or not r is statistically different from zero. At the point where r cannot be distinguished from zero, we define the *range* of the variable, that is, the average distance over which the variable shows autocorrelation. One tests for anisotropy by comparing ranges obtained for different directions. Three examples discussed by Agterberg (1965) suggest that it may not be very critical whether data are logtransformed or not for purposes of studying autocorrelation. However, it seems likely that these three test cases are not representative, and as a rule the form of histograms should be evaluated to determine if log transformation is desirable *prior* to examining autocorrelation. Hodgson (1972) has discussed the practical applications of these techniques to regional soil surveys.

Covariograms or semivariograms can be used for comparable studies (e.g., David and Dagbert, 1975).

Possible problems in correlation studies

(1) Departures of the variables from normality destroy the ability to test correlation coefficients for significance. Outliers are a particular cause of departures from normality—a single outlying value can render a calculated correlation coefficient meaningless.
(2) Nonlinear trends are common in geochemical data and may go unrecognized if x–y plots of all pairs of variables are not examined.
(3) The meaning of correlations among closed variables (percentages or proportions that sum to 100% or 1, respectively) is uncertain. The problem may decrease as the number of variables involved in the sum increases.
(4) Ratios of one element to another are difficult to interpret—they should be avoided where statistical analysis is to be done. The correlation between two ratios is a function of the various paired correlations and the coefficients of variation of the 4 variables involved in the ratios.
(5) Autocorrelation methods are used to quantify spatial correlation. They are important in defining ranges which can be considered an optimum sample spacing. In soil surveys oriented towards mineral exploration the range may represent the average dimension of an anomalous zone and consequently can be used to determine "optimal" sample spacing, probability of obtaining two samples in a specified size of anomalous zone, and so on. Such calculations are highly idealized!

SIMPLE LINEAR REGRESSION

Introduction

There are many practical situations in which it is desirable to fit a straight line to a set of paired data. In fact, we have seen how the simple linear correlation coefficient is a measure of the extent to which a straight line pattern exists in a set of data. Having established a nonrandom distribution of points on a scatter diagram (by the existence of a linear correlation coefficient that is significantly different from zero) it is a simple matter to calculate the equation of a straight line as a first approximation to describing the nonrandomness of the data.

The general procedure requires two normally distributed variables x and y, with y being the dependent variable and x the independent variable by conventional definition. We will define the "best fit" straight line as the line about which there is a minimum variance (i.e. the least squares line).

An equation for such a straight line takes the form

$$y = b_0 + b_1 x + e$$

or

$$\sum y_i = nb_0 + b_1 \sum x_i + \sum e_i$$

The "best fit" straight line must minimize

$$\sum e_i^2 = \sum (y_i - b_0 - b_1 x_i)^2$$

If the right hand side is differentiated twice, once with respect to b_0 and once with respect to b_1, and both results equated to zero, one obtains the *normal equations*

$$\sum y_i - nb_0 - b_1 \sum x_i = 0$$
$$\sum y_i x_i - b_0 \sum x_i - b_1 \sum x_i^2 = 0$$

All summations are known from a given data set and the normal equations can be solved for b_0 and b_1. By appropriate rearrangement the solution can be expressed in terms of sums of squares (see definitions in section on correlation).

Summary of formulae

(1) $$b_1 = \frac{\sum(y_i - \bar{y})(x_i - \bar{x})}{\sum(x_i - \bar{x})^2}$$
$$= \frac{\sum(y_i x_i) - (\sum y_i)(\sum x_i)/n}{\sum x_i^2 - (\sum x_i)^2/n}$$
$$= \frac{SS_{xy}}{SS_x}$$

(2) $$b_0 = \bar{y} - b_1 \bar{x}$$

(3) The prediction of y for any value of x is given by
$$y = b_0 + b_1 x$$

(4) $$\sigma^2 = \frac{1}{n-2}\sum(y_i - b_0 - b_1 x_i)^2$$
$$= \frac{1}{n-2}\sum(y_i - \bar{y})^2 - \frac{[\sum(y_i - \bar{y})(x_i - \bar{x})]^2}{\sum(x_i - \bar{x})^2}$$
$$= \frac{1}{n-2}\left[SS_y - \frac{(SS_{xy})^2}{SS_x}\right]$$

(5) The confidence interval on b_0 with confidence $1 - \alpha$ is
$$\pm\left[t_{\alpha/2}(n-2)\sqrt{\frac{\sigma^2 \sum x_i^2}{n(SS_x)}}\right]$$

(6) The confidence interval on b_1 with confidence $1 - \alpha$ is
$$\pm\left[t_{\alpha/2}(n-2)\sqrt{\frac{\sigma^2}{n(SS_x)}}\right]$$

Note that two different least squares equations are obtained depending on which variable is taken as the dependent variable |y|. This is implicit in the estimate of slope (b_1) where the value of the denominator (SS_x) will vary depending on which variable is taken as x. The straight lines calculated in this manner pass through the mean values of x and y.

Confidence limits on b_0 and b_1 may be very important. For example, b_0, the y-intercept, might be tested to see if it is statistically distinguishable from zero. This may be a fundamental test in establishing the presence of bias in duplicate samples analyzed by two laboratories or by different procedures. Similarly, b_1 may be compared with the slope of another line.

Some applications of linear regression

Common application of linear regression include:

(1) Establishment of working curves.
(2) Quality control in geochemical surveys.
(3) Generalization of simple trends.
(4) Estimation of variables that are costly or difficult to obtain, by other variables that are relatively easy or cheaper to determine; e.g.,
   (a) Specific gravity may be used to estimate Fe content of mineralogically simple iron formation.
   (b) Scintillometer response can be equated to uranium equivalents.
(5) The simple linear model described above is substantially more general than might appear. Providing x and y are completely determinable from available data the linear model applies even though a variety of functions might be involved, as follows:

$$y = b_0 + b_1 \log Z \pm e$$
$$y = b_1 Z^2 \pm e$$
$$y = b_0 + b_1 e^{-Z} \pm e$$
$$y = b_0 + b_1 \cos 3Z \pm e$$

Scatter diagrams are plots of the two variables that can be used to show graphically how well a straight line fits a particular data set. More rigorously, the correlation coefficient can be tested for significance at various levels by reference to tabulations for various degrees of freedom.

A study of characteristics of sediments in a tidal flat (Mud Bay) near Vancouver, B.C., has been reported by White and Northcote (1962). One hundred and thirteen pairs of values of "% minus 200 mesh" and "% sulfur", both with approximately normal distributions have a high correlation coefficient (r = 0.682). These data are shown in Figure 5.14 where the nonrandom disposition of data is apparent. A least squares line has been fitted to the data by regressing "% minus 200 mesh" on "% sulfur". The result is given by equation (1) in Table 5.6 and shown in Figure 5.14. It is apparent that this line overestimates low values of "% minus 200 mesh" for a given sulfur percentage (and underestimates high values) due to the pronounced weighting imposed on the calculations by the two outliers.

It may be more desirable to regress "% sulfur" on "% minus 200 mesh". This has been done *omitting the two out-*

TABLE 5.6—Simple linear regression equations and estimators of population parameters, Mud Bay. Sulfur and size fraction (minus 200 mesh) data.

Equation 1	(% −200 mesh) = 52.0(%S_2) − 4.37	
%S_2 = 0.766		(% −200 mesh) = 35.4%
std. dev. = 0.305		std. dev. = 23.2
r^2 = 0.466		
r = 0.682		
n = 113		
Equation 2	(%S_2) = 0.008(% −200 mesh) + 0.471	
%S_2 = 0.742		(% −200 mesh) = 33.8%
std. dev. = 0.247		std. dev. = 22.5
r^2 = 0.527		
r = 0.726		
n = 111		

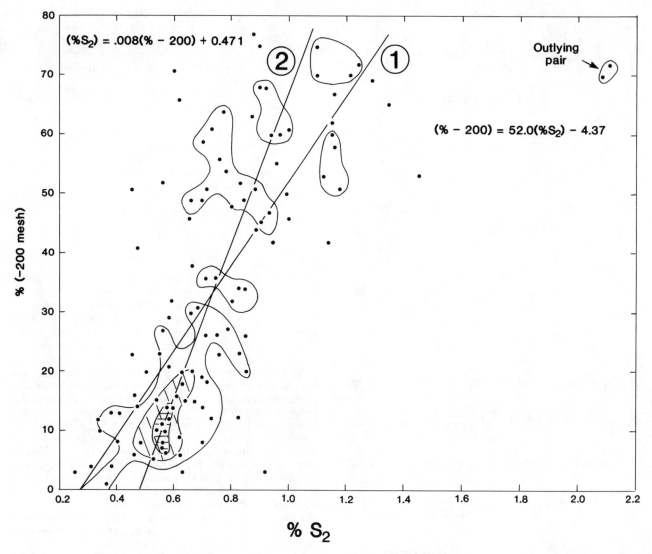

FIGURE 5.14—Scatter diagram of 113 sediment samples from Mud Bay (a tidal flat), weight percent of minus 200 mesh size fraction versus weight percent sulfur of that size fraction. Black dots are original data. Contours are at the 1, 2 and 5 samples per unit area (see text). Note 2 outlying values in upper right of diagram. See text and Table 5.6 for an explanation of straight lines 1 and 2 fitted to the data.

lying values to which reference has been made, and results are listed in Table 5.6 as equation (2) and are shown in Figure 5.14. Line 2 obviously "describes" the great bulk of the data better than does line 1 emphasizing the importance of examining scatter diagrams to check for the influence of outliers. Note that with omission of the two outlying values the correlation coefficient has improved slightly. In some cases, particularly where small data sets are concerned, the removal of outliers can produce a dramatic change in the correlation coefficients.

Contouring scatter diagrams is a useful way of examining groupings and trends in samples and is an interesting approach to bivariate classification. A rectangular grid with cell size "0.1% S_2" by "5% minus 200 mesh" was used as a basis for contouring (Figure 5.14). The number of points in each cell was counted, attributed to the cell center and the resulting regular grid contoured by linear interpolation with slight aesthetic smoothing of contours. Contours are at 2, 3 and 5 points per cell. In this case the general trend of highs is expressed best by equation (2). In general, contouring of scatter diagrams is a useful data evaluation scheme that may lead to the recognition of multiple trends or classes that otherwise are not easily recognizable.

Degree of fit

The *coefficient of determination* is the square of the simple correlation coefficient and gives the proportion of total variability "explained" by the straight line. As a percentage this becomes 100 r^2.

Errors in both variables

The procedures discussed to date assume the error to be entirely in the dependent variable. As a rule, however, in geochemical examples a substantial error exists in both dependent and independent variables. Mark and Church (1977) have considered this problem and provide the following equation for slope to take such error situations into account.

$$\hat{b}_1 = \frac{SS_y - \lambda SS_x + \sqrt{(SS_y - \lambda SS_x)^2 + 4\lambda(SS_{xy})^2}}{2SS_{xy}}$$

where
$$\lambda = \frac{\sigma^2(y)}{\sigma^2(x)}$$

In geochemical applications $\sigma^2(y)$ and $\sigma^2(x)$ represent the inherent local variations plus analytical errors in samples. Where these errors are not known with confidence they are in many cases assumed to be proportional to variances of the variables to produce what is referred to as a "Reduced Major Axis" solution. In cases where errors in the two variables are almost the same, the assumption that $\lambda = 1$ provides the so-called "Principal Axis" solution.

CHI SQUARE DISTRIBUTION

Introduction

The chi square test is non-parametric in nature, meaning that its application is not dependent on a particular probability density function for the variables being tested. In fact, one of its principal applications is to test whether or not a sample might have been drawn from a population with a pdf of a particular form. The second important application is in the evaluation of 2-way contingency tables, especially in testing the variables in such tables for dependence or independence.

Goodness of fit

Consider the histogram in Figure 5.2b. We wish to question whether or not the data are described adequately by a normal distribution. The chi square test involves a measure of "departures from expected values", and provides a means of comparing measured departures with statistically expected departures utilizing the following statistic

$$U = \sum_{i=1}^{n} \frac{(O_i - E_i)^2}{E_i}$$

which has a chi square distribution with $n-p-1$ degrees of freedom. P is the number of parameters involved in the null hypothesis. In the example cited in Figure 5.2b, the frequency of each class interval can be compared with the expected frequency of a normal distribution having the same mean and variance as the data on which the histogram is based. Such a normal curve has been fitted to the histogram although this is not necessary in conducting the test. The observed (O_i) and expected (E_i) values are compared in Table 5.7. Note that it is particularly convenient in conducting this test manually to have the class interval precisely one quarter the standard deviation and to have the classes distributed symmetrically about the mean values. If constructed in this fashion areas under the normal curve corresponding to each class interval are particularly easy to estimate and compare with the corresponding frequency (as a percent). Where gaps exist (class interval with zero frequency) the classes must be grouped. Not more than 20 percent of the classes should contain an absolute frequency less than 5.

In the example of Figure 5.2b and Table 5.7 there are 6 classes and therefore $6 - (2+1) = 3$ degrees of freedom. There are two fewer degrees of freedom than would be the case if the mean and standard deviation of the distribution were known rather than being estimated from the histogram. The critical chi square value at the 0.05 level and 3 degrees of freedom is 7.81. The calculated value in Table 5.7 is 3.73 and we conclude that the density (histogram) of

TABLE 5.7—Normal curve compared with histogram using chi square – goodness of fit. Hg in B horizon soils.*

Lower class interval		Frequency		
Absolute	Z-score	Expected (E_i)	Observed (O_i)	$(E_i - O_i)^2/E_i$
55	2.012	2.22	1.7	0.1218
45	1.229	9.09	12.1	0.9967
35	0.446	21.51	18.1	0.5406
25	-0.338	30.46	26.7	0.4641
15	-1.121	23.40	28.4	1.0684
5	-1.904	10.52	12.9	0.5384
				3.7300**

*Same data as Figure 5.2b and Table 5.2
**d.f. = 6 - (2+1) = 3
chi square (α = 0.05, d.f. = 3) = 7.81

TABLE 5.8—Two-way contingency table for 82 soil samples classed with respect to B horizon color and parent material.

Parent material	B horizon color			Row totals
	Brown	Yellow-brown	Yellow	
Coarse till	6(5.4)*	7(6.0)	4(5.5)	17
Medium till	9(6.7)	5(7.4)	7(6.9)	21
Fine till	6(6.3)	8(7.1)	6(6.6)	20
Varved clay	5(7.6)	9(8.5)	10(8.0)	24
Column totals	26	29	27	82

*Observed count (expected count)

log transformed data does not depart substantially from a normal distribution. A simple example has been used to demonstrate the goodness of fit technique which is crude in this case because number of classes is so low (less than 10).

A comparable approach can be used for comparing a histogram with other forms of probability density functions. The test in many applied geochemical applications is not particularly discriminating. In fact, it is not unusual that a data set can be shown to fit both normal and lognormal probability density functions.

Two-way contingency tables

An interesting and relatively little used chi square procedure involves comparing the distribution of items among different categories of two variables. Consider a two-way table such as that shown in Table 5.8 where B horizon soil samples have been classed on the basis of underlying glacial sediment class and color of the B horizon. The table meets the requirements of the chi square test.

Expected values can be determined on the assumption of independence; that is, the null hypothesis states that the same general distribution of variable one groupings exists regardless of which variable two category is considered. For example, if paired sediment type is independent of soil color the distribution of values in all rows of Table 5.8 should be more or less in the same ratio. Expected values are determined by the relation

$$E_{ij} = \frac{\sum R_i \sum C_j}{\sum O_{ij}}$$

where E_{ij} is the expected value for the slot at the intersection of row i and column j.
ΣO_{ij} is the grand sum
ΣR_i is the sum of row i
ΣC_j is the sum of column j
O_{ij} is the observed value at the intersection of row i and column j.

Expected values are the bracketed figures in Table 5.8.

It is now a simple matter to conduct a chi square test based on differences between all O_{ij} and E_{ij}, where

$$U = \sum \frac{(E_{ij} - O_{ij})^2}{E_{ij}}$$

If U is greater than a critical chi square value for a predetermined level of significance (say 0.05) and known degrees of freedom $(n_r - 1)(n_c - 1)$, the differences are sufficient to indicate that the two variables are dependent, that is, the distribution of one depends on the other. Conversely, if U is less than the critical chi square value, the distribution of one variable is the same for all categories of the second variable. For the example cited here (Table 5.9) U = 3.27, compared with a critical value of 12.59, and the two variables are interpreted as independent, that is, the distribution of B horizon soil colors is essentially the same regardless of the parent glacial sediment.

TABLE 5.9—U-values for two-way contingency Table 5.8.

Parent material	B horizon color		
	Brown	Yellow-brown	Yellow
Coarse till	0.067	0.167	0.400
Medium till	0.790	0.798	0.001
Fine till	0.014	0.114	0.054
Varved clay	0.336	0.030	0.500
Grand total (U)			3.271

d.f. = (4−1)(3−1) = 6
chi square (α = 0.05, d.f. = 6) = 12.59

FINAL REMARKS

The philosophy of the preceding pages emphasizes relatively simple data analysis techniques followed by progressively working towards more complicated procedures as a particular case requires. In many practical cases sophisticated statistical methods are not essential and add little to the understanding of a geochemical data set. Those intending a serious application of statistical procedures should ensure first and foremost that they understand the statistics involved and then consider the limitations in application to real data. Apart from numerous statistical texts readers are referred to Howarth (1983) as an up-to-date source of information and references on quantitative approaches to interpreting geochemical data.

REFERENCES

Agterberg, F.P. 1965. The technique of serial correlation applied to continuous series of element concentration values in homogenous rocks. Journal Geology, v. 72, p. 142–154.

Bolviken, B. 1971. A statistical approach to the problem of interpretation in geochemical prospecting. Canadian Institute Mining and Metallurgy, Special Volume 11, p. 564–567.

David, M. and Dagbert, M. 1975. Lakeview revisited: variograms and correspondence analysis—new tools for the understanding of geochemical data. In: Elliott, I. and Fletcher, W.K. (editors), Geochemical Exploration 1974, Developments in Economic Geology Volume 1. Elsevier Scientific Publishing, 720 pp.

Godwin, C.I. and Sinclair, A.J. 1979. Application of multiple regression analysis to drill target selection, Casino porphyry copper-molybdenum deposit, Yukon Territory, Canada. Transactions Institute of Mining and Metallurgy, Section B, v. 88, p. 93–106.

Harris, S.A. 1958. Probability curves and the recognition of adjustment to depositional environment. Journal Sedimentary Petrology, v. 28, p. 151–163

Hodgson, W.A. 1972. Optimum spacing for soil sample traverses. Proclamations 10th APCOM Symposium, South African Institute Mining and Metallurgy, Johannesburg, p. 75–78.

Howarth, R.J. (editor) 1983. Statistics and Data Analysis in Geochemical Prospecting. Handbook of Exploration Geochemistry, Volume 2. Elsevier Scientific Publishing, 437 pp.

Krumbein, W.C. and Graybill, F.A. 1965. An Introduction to Statistical Models in Geology. McGraw–Hill Publishing Co., New York, 475 pp.

Lepeltier, C. 1969. A simplified statistical treatment of geochemical data by graphical representation. Economic Geology, v. 64, p. 538–550.

Mark, D.M. and Church, M. 1977. On the misuse of regression in the earth sciences. Mathematical Geology, v. 9, no. 1, p. 63–75.

Matheron, G. 1963. Principles of geostatistics. Economic Geology, v. 58, pp. 1246–1266.

McCammon, R.B. and Wenninger, G. 1970. The dendrograph. Computer contribution 48, State Geological Survey, The University of Kansas, Lawrence, 17 pp.

Montgomery, J.H., Cochrane, D.R. and Sinclair, A.J. 1975. Discovery and exploration of Ashnola porphyry copper deposit near Keremeos, B.C.; a geochemical case history. In: Fletcher, W.K. and Elliott, I. (editors), Geochemical Exploration 1974. Elsevier Pub. Co., Amsterdam, pp. 85–100.

Parslow, G.R., 1974. Determination of background and threshold in exploration geochemistry. Journal Geochemical Exploration, v. 3, p. 319–336

Saager, R. and Sinclair, A.J. 1974. Factor analysis of stream sediment geochemical data from the Mount Nansen area, Yukon Territory, Canada. Mineralium Deposita, v. 9, p. 243–252.

Shaw, D.M. 1961. Element distribution laws in geochemistry. Geochimica et Cosmochimica Acta, v. 23, p. 116–124.

Sinclair, A.J. 1974a. Selection of thresholds in geochemical data using probability graphs. Journal Geochemical Exploration, v. 3, p. 129–149.

Sinclair, A.J. 1974b. Probability graphs of ore tonnage in mining camps—a guide to exploration. Bulletin Canadian Institute Mining and Metallurgy, v. 67, p. 71–75.

Sinclair, A.J. 1976. Application of Probability Graphs in Mineral Exploration. The Association of Exploration Geochemists, Special Volume 4, 95 pp.

Stanley, C.R. 1984. The geology and geochemistry of the Daisy Creek Prospect, a stratabound copper-silver occurrence in western Montana. M.Sc. thesis. Department of Geological Sciences, University of British Columbia, Vancouver, 277 pp. plus maps.

White, W.K., and Northcote, K.E. 1962. Distribution of metals in a modern marine environment. Economic Geology, v. 57, p. 405–409.

Chapter 6

Models, Interpretation and Followup

S. J. Hoffman and I. Thomson

MODELS

General Background

Idealized or conceptual models are now so commonly encountered in economic geology that it is hardly necessary to define the principles involved in their construction and application. Nevertheless, it is useful to explore in a little more detail the background and development of the exploration geochemistry models, an example of which is shown in Figure 6.1, that are now in use around the world.

During the early 1970's a small group of professional geochemists working in Canada came to realize that sufficient common experience existed to make general conclusions as to the mechanisms of geochemical dispersion and the morphology of geochemical anomaly patterns that could be anticipated in any given situation. Furthermore these features could be classified according to certain key features and thus simplified into a series of idealized models. The result was a major publication, edited by Bradshaw (1975), entitled Conceptual Models in Exploration Geochemistry: The Canadian Cordillera and Canadian Shield. This was followed by similar compilations for Norden (Kauranne, 1976), the Basin and Range Province of the Western United States and Northern Mexico (Lovering and McCarthy, 1978) and Australia (Butt and Smith, 1980).

The rationale for this work is well summarized by Bradshaw (1975) who noted that explorationists "find considerable difficulty in absorbing the large volume of data which exists on exploration geochemistry... It is difficult to draw together the large number of complete and partial case histories which exist in the literature and to obtain valid conclusions and generalizations from these data. Although individual case histories are vital to understanding the mechanisms controlling geochemical dispersion they are in fact a laborious approach to understanding. For situations are so varied that even a number of case histories might be a misleading example, whilst each is so complex that even a detailed description may be too summary: and none is comprehensible outside of its context." The models are, therefore, a synthesis of existing data and provide an overall understanding of the mechanisms of geochemical dispersion and a framework into which further data can be fitted or upon which data may be interpreted.

Landscape Geochemistry

The conceptual models, as presented here and in the publications noted above, are themselves constructed within the framework of landscape geochemistry.

(For a complete discussion of landscape geochemistry, which originated in Russia during the 1930's, you are referred to the useful textbook by Fortescue (1980)). Landscape geochemistry is a holistic approach that involves consideration of the complete environment. Landscape is here defined as a dynamic system involving the relationships between vegetation, soils, underlying rocks, the atmosphere, surface and groundwaters, geomorphology and geology. It is these interrelationships at or near the daylight surface that govern the migration (dispersion) of elements.

Fortescue (1980) identifies six fundamental concepts within landscape geochemistry. These are:

(1) Element abundances—the absolute or relative (partial, selectively extractable, etc.) abundance of elements in a given medium.
(2) Element migration—the movement of elements, their absolute and relative mobility and the forms in which movement takes place.
(3) Geochemical flow—the pathways or plumbing systems along which element migration takes place and the speed at which this proceeds.
(4) Geochemical gradients—the rate of change in the abundance of elements. This is often descriptive of changes in substrate, geochemical flow and geochemical barriers.
(5) Geochemical barriers—these are caused by changes in conditions usually related to migration (Eh, pH, etc.) or flow (permeability, porosity, etc.)
(6) Historical development—the position in time in the evolution of the landscape such as partial or complete development of a process with a defined end point (e.g., podzolization of soil), overprinting by a change of conditions, pollution, contamination, etc.

At this point it is worth pausing to consider these concepts a little more fully. Refer to a recent exploration case history from your own experience and

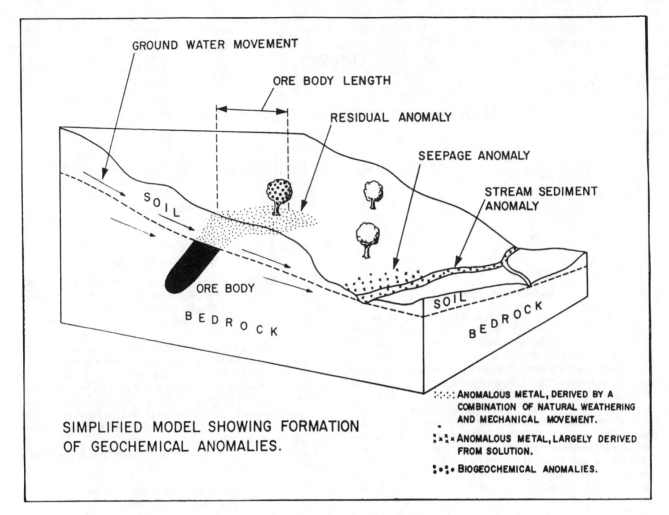

FIGURE 6.1—Simplified model showing formation of geochemical anomalies.

try to categorize the geochemical environment and survey data in terms of the six fundamentals. Alternatively list a series of examples from your experience or from the literature that illustrate each of the fundamental concepts.

From these fundamentals it is an easy step to consider geochemical data as an expression of landscape. By recognizing that patterns of element abundances are an expression of changes in a given medium, geochemical barriers, gradients, flow and migration, it is possible to interpret a landscape and identify the underlying controls. The exploration geochemist can be more specific and focus on those features that permit recognition of the presence of potentially economic mineral deposits and may be used to locate the site of the mineralization.

Idealized Models

Idealized models (Figure 6.1) represent pictorially the general conclusions (both positive and negative from the point of view of economic application) concerning the dispersion of metals from mineralizations and the formation of anomaly patterns in any given landscape configuration. Every model has certain common features.

(1) A body of mineralization or a rock type, etc. that may mimic mineralization.
(2) The relative distribution of bedrock, overburden, soil, groundwater, surface water, vegetation, etc.
(3) Dispersion pathways related to mineralization and anomaly formation are highlighted.

Three types of three dimensional diagrams are used to illustrate both the broad and local (detailed) condition: (a) idealized models—which show the total environment about a geochemical anomaly, (b) idealized cross sections—which display geochemical characteristics on a continuous section, and (c) idealized prisms—which show details of vertical changes within a particular profile. In certain cases the cross sections and prisms may be exploded to show more detail.

In constructing or using idealized models it is important to remember that they have no scale. They indicate the mechanisms of formation of geochemical patterns only and not magnitude or size. The mechanisms (mechanical or hydromorphic dispersion, etc.) are fundamental to a variety of situations, while dimension and relief of geochemical patterns are influenced by many local conditions and cannot be summarized.

Finally it must be realized that at least two types of model may be developed. It is fundamental to the published compilations of case histories and models listed above that no model, nor any aspect of the models, is drawn without the support of field examples. These are empirical models based on experience. An alternate form of model can be, and frequently is, drawn, often as an aid to interpretation or planning, which is based entirely upon assumptions of what might be happening. These theoretical models can be very useful but must be used with caution until they are supported by real data.

The empirical models are the first step towards full dynamic modelling and simulation of dispersion. Work in this area will probably lead to future, higher levels of sophistication in exploration geochemistry. Current research in low temperature thermodynamics and studies on the stability fields of secondary minerals, carried out in support of programs in radioactive waste disposal and environmental science, is providing the necessary data for computer simulations that can be used in mineral exploration. However, while this will provide for greater confidence in interpreting processes, other features of the landscape (overburden, topography, etc.) are likely to remain so infinitely variable that they will defy reliable simulation for some considerable time.

Examples

Many models have been constructed and published to illustrate situations encountered around the world. It is quite impossible to present them all at this time so only a selected

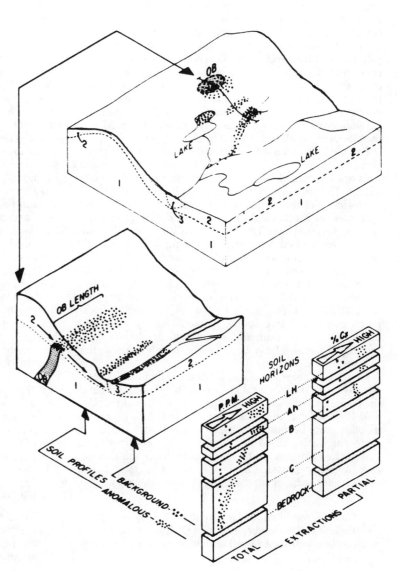

FIGURE 6.2—Idealized models for geochemical dispersion of mobile elements in well and poorly drained residual soils. After Bradshaw (1975).

series will be considered briefly. For more complete information and specific models you are referred to the series of compilations published in the Journal of Geochemical Exploration (Bradshaw, 1975; Kauranne, 1976; Lovering and McCarthy, 1978; Butt and Smith, 1980).

The simplest situation ordinarily encountered in exploration is shown in Figure 6.2. This model depicts geochemical dispersion of mobile elements from mineralization in a landscape characterized by residual soils, which are freely drained except in seepage areas and bogs, and modest youthful topography. In this situation geochemical anomalies are formed directly over mineralization during the normal processes of soil formation involving both mechanical and chemical modification. The anomaly may be roughly the dimensions of the subcropping mineralization. More commonly, lateral spreading due to downslope creep or related to soil compaction will result in an anomaly with larger dimensions than the underlying mineralization. This modification is dominantly the result of mechanical dispersion processes. In addition to soil-forming processes, metal is taken into solution in the acidic environment of weathering sulfide mineralization close to the water table. This metal moves with the groundwater and remains in solution until a change in the chemical environment is encountered. Such a change occurs where groundwaters enter the oxidizing and generally less acidic conditions of the surface environment either in seepage areas at the break of slopes or in lakes and streams. This hydromorphic dispersion gives rise to secondary anomalies displaced from the site of mineralization in bedrock.

The model in Figure 6.3a illustrates dispersion of mobile elements from mineralization at a site where well drained soils are developed on glacial till of essentially local derivation. The situation is basically a modification of the first model in which the surface soil anomaly feature is distorted, elongated by mechanical smearing in a down ice direction due to glacial action. The soil anomaly is typically much larger than the bedrock source becoming disrupted or diffuse due to mechanical dilution from unmineralized material incorporated during ice transport. Hydromorphic dispersion processes are essentially similar to the first case. However, chemical weathering of mechanically dispersed mineralized rock fragments in the till provide an areally much more extensive source of metal for hydromorphic dispersion. As a result the secondary, displaced hydromorphic anomalies may be similarly more extensive than in a residual environment.

Figure 6.3b shows the same block model with soils developed on glacial till of local origin. In this model, however, the elements dispersed from mineralization are immobile—lead from galena, tin in cassiterite or free gold grains—or mobility may be inhibited by local environmental conditions—high pH caused by carbonate rock, etc. In this situation the elements are transported by mechanical means only forming a down ice dispersion fan due to glacial smearing further modified, perhaps, by downslope creep.

Can you modify these last two drawings by placing the mineralization beneath a bog? How do the two models, with mobile and immobile elements, differ now?

Further modifications to these models occur as a function of drainage conditions, relief and overburden. In respect of the latter, the presence of thick till or exotic glacial sediments, notably stratified drift, will prevent the development of mechanically derived anomalies in soil over or adjacent to mineralization (Figure 6.4). Under such circumstances other dispersion processes become important in delivering evidence of the presence of mineralization to the daylight surface.

The latter situation is not confined to glaciated terrain. Any overburden of exotic origin, (alluvium, volcanic ash, lake clay, etc.) will act in this way rendering the model shown in Figure 6.4 applicable throughout the world simply by changing the name of the transported overburden.

Mineralization is not the only geological source of variability in the geochemical landscape. As shown in Figure 6.5, wherever a rock type with a high metal content occurs it will give rise to distinctive geochemical patterns in soils (and sediments) that may on occasion mimic mineralization. This is due to the normal weathering of bedrock, which results in trace and major elements being incorporated in the overlying soil. Hydromorphic dispersion away from rock units with a high metal content is usually much less than for mineralization because of the relative stability of rock-forming minerals and the absence of sulfides to lower the pH of the environment. The model drawn in Figure 6.5 relates to areas with residual soils. However, the same modifying effects described earlier for glaciated environments and exotic overburden can be applied to this situation and a model descriptive of each case drawn with relative ease.

The four examples presented here represent end member situations. As local conditions change, so the models may be modified or new models constructed. In practice the most important local variables are element mobility, dispersion process, surface soil conditions, overburden thickness and overburden composition. To this may be added landform, climate and the weathering history of an area.

Considerable complexity may be accommodated by the models as in the case of the Australian compilation (Butt and Smith, 1980). Figure 6.6 is a model illustrating the character of geochemical dispersion patterns in a deeply weathered environment of moderate relief. Of note is the fact that, in Australia, these deeply weathered profiles developed under humid climatic conditions that prevailed in the Mesozoic and early Tertiary. They have been preserved because of the tectonic stability of the Australian continent and the trend towards aridity that persists today.

> Turn your mind to the last exploration survey you were involved with. Draw an idealized model of the property placing the mineralization in the center and taking care to include as much as you know of overburden conditions, groundwater and soils. Finally, sketch in geochemical dispersion patterns. Can you interpret what is going on in terms of Landscape Geochemistry? What does the model tell you of your knowledge of the survey area?

> Refer to the orientation study from Norway in Chapter 1. Read the background information and carefully examine the tables and figures. Then summarize the orientation work by drawing an

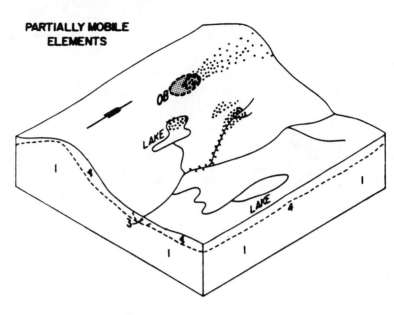

FIGURE 6.3—Idealized models of the effect of chemical mobility of elements on their dispersion pattern in till covered areas. After Bradshaw (1975).

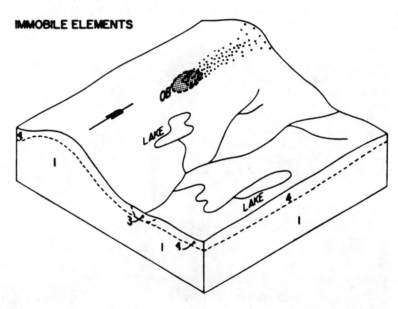

idealized block model that describes the landscape of the survey area. What is the composition of the overburden in your model, and what is the dominant mode of dispersion of base metals from mineralization?

Applications

Idealized models have several important applications, the most important of which are:

(1) In the design of appropriate exploration procedures, particularly the choice of sample medium and sampling strategy.

(2) In providing a framework for the interpretation of survey data.

(3) In assisting communications by summarizing a large amount of complex information into a form that is comprehensible to both specialists and nonspecialists and workers in other disciplines. Models are particularly useful as teaching aids.

(4) As a means of rapidly recognizing those areas where information is not available and thus prompting completion of a survey, further orientation studies or necessary research work.

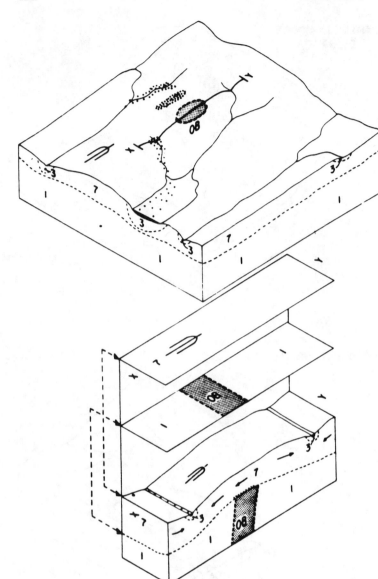

FIGURE 6.4—Idealized models for geochemical dispersion of mobile elements in areas of stratified drift. After Bradshaw (1975).

INTERPRETATION

A typical, but poorly conceived, approach to interpretation is illustrated on the left hand side of the Interpretational Flow Chart, Part 1 (Figure 6.7). With this approach the raw data are manually scanned and plotted to obtain a "feel" for the numbers. Values are then contoured using "appropriate" but subjectively selected, constant-interval increments of concentration to indicate areas worthy of followup. Multielement data may be available but only supposedly "key" elements are plotted. Such a procedure may highlight the most obvious of anomalies (real or false) but is unlikely to be truly effective insofar as it ignores (a) geological influences on metal distributions and (b) the basic principles of the geochemical models described in the preceeding section. As a result undue importance may be attached to false anomalies, while the significance of subtle anomalies related to mineralization goes unrecognized.

In contrast to the foregoing approach, the right hand side of the Flow Chart (Figure 6.7) represents the first steps in an integrated scheme that continues on into three more Flow Charts (Figures 6.8, 6.9 and 6.10). The thought processes guiding interpretation are described below emphasizing why, at each step, the interpreter has to answer fundamental questions.

Interpretation begins with the arrival of geochemical data listings from the laboratory. Histograms and/or probability plots are drawn and the data portrayed in the best fashion to describe the trace element distribution on the property. Class intervals and scales are selected to give normal or lognormal distributions after the very highest values are removed to eliminate their influence on frequency diagram

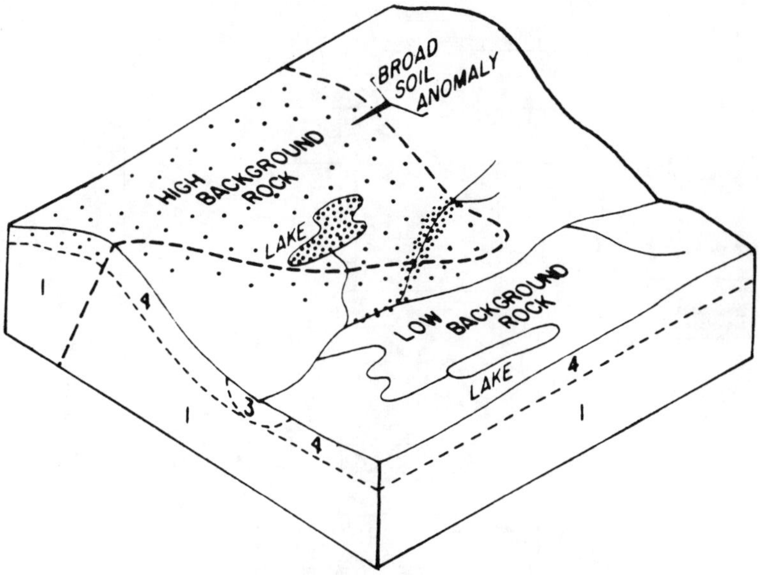

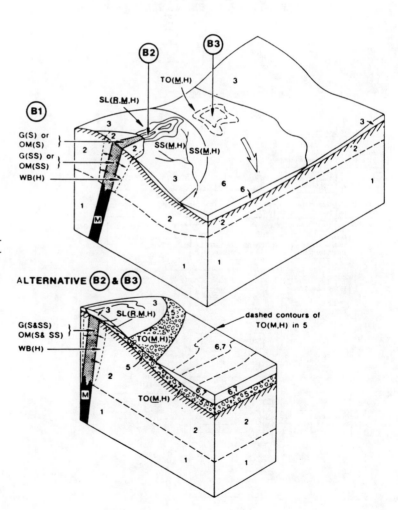

FIGURE 6.5—Idealized model for the effect of rock type change on the geochemistry of the overlying soils and sediments. After Bradshaw (1975).

FIGURE 6.6—Idealized model of geochemical dispersion in a deeply weathered environment with moderate relief, Australia. After Butt and Smith (1980).

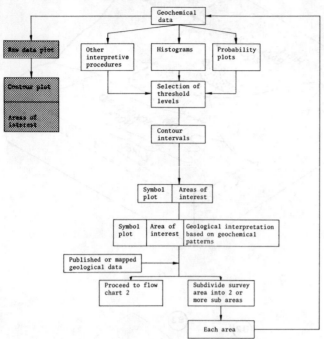

FIGURE 6.7—Interpretational flow chart—Part 1.

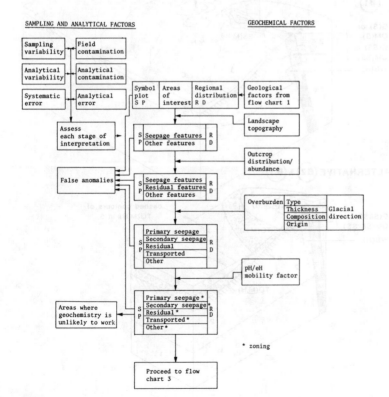

FIGURE 6.8—Interpretational flow chart—Part 2.

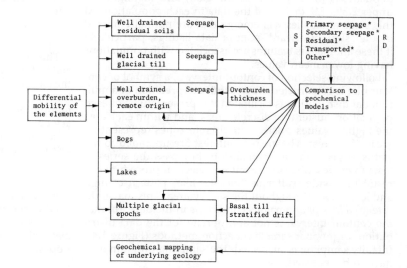

FIGURE 6.9—Interpretational flow chart—Part 3.

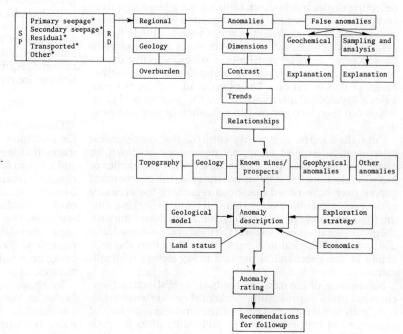

FIGURE 6.10—Interpretational flow chart—Part 4.

shape. These high values represent the "sore thumb" geochemical features and are often the only anomalies that would be recognized by the interpretive procedures on the left hand side of Flow Chart 1. For the remaining data, contour intervals are chosen to clearly differentiate multiple populations (if they exist in the data) and highlight the upper 2.5%, 5% or 10% of the data in each population. At this stage anomalies will emerge that are normally missed by the "straightforward" scan of data listings, particularly if these anomalous features are associated with populations having lower mean values.

Following selection of contour intervals, a symbol or contour plot is constructed. Successful interpretation of a multimodal distribution is facilitated if representatives of each population cluster into discrete areas, and within each area the highest values cluster into discrete zones representing areas of interest. The areas of interest are upgraded to the status of geochemical anomalies if they pass the series of tests to be described. The contour selection procedure can result in a wide scatter or "measles-like" pattern of high and low values throughout the survey area. This may indicate (a) a lack of anomalous conditions in the data, (b) that the contour interval is not appropriate for the data distribution, or (c) poor sample collection methods (Figure 1.4). If faulty sampling is suspected, the sampling program may have to be repeated.

Providing sampling was satisfactory and is not a major source of variability, the geochemical maps are examined for evidence of regional controls on metal levels exerted by underlying geology or distinctive overburden types. Association with the latter should be apparent if geochemical results are compared to topographic information and field notes as described in Chapter 3. Geochemical features reflecting distinctive bedrock units (or very large alteration and primary geochemical halos associated with mineralized zones) tend to be multisample zones with relatively homogeneous metal values. Geological controls are therefore to be suspected if metal distribution zones, varying over a narrow range of concentrations, abruptly change to another range of concentrations. Distribution of such zones provides a "geological interpretation" of the geochemical data, which can then be compared to published or mapped geology.

Published maps commonly confirm the geochemical interpretation (subject to normal downslope or down ice dispersion)—for example, Cr associated with ultramafics or Mo with black shales. If this is not the case, the geochemical survey may have raised questions regarding the accuracy of the geological information that will require checking during followup surveys. Conversely, failure to detect the geochemical signature of a well-established and geochemically distinctive lithological unit must throw doubt on the suitability of the geochemical method for detecting mineralization.

Subdivision of the data, on the basis of distinctive "geochemical units" representing geological or overburden factors, leads to the construction and interpretation of a second generation of histograms and/or probability plots for each subset. This can be repeated, as often as necessary, to eliminate broad scale regional patterns and focus on more local geochemical features.

Flow Chart—Part 2 (Figure 6.8) begins with a symbol plot (SP) on which are outlined "areas of interest," indicated by high values relative to adjacent lower values, and regional distributions related to geological features as discussed above. "Areas of interest" are then considered further in relation to the factors shown on the right hand side of the chart, namely:

Landscape/Topography

This is viewed in relation to its favorability or unfavorability towards geochemical dispersion and its control on the location of seepage anomalies. Generally, steeper slopes favor active mechanical and hydromorphic dispersion, promoting formation of geochemical anomalies in the secondary environment. Caution should be taken in interpreting geochemical features where topography is associated with distinct changes in overburden type.

Outcrop

Abundant outcrops of bedrock suggest a relatively thin overburden in adjacent areas through which geochemical dispersion might be active. They can also be a source of material shed mechanically (colluvium) onto surrounding slopes.

Overburden

Thickness and type of overburden are considered next. Different types of overburden are recognized based on their compositional and/or landform characteristics as described in Chapter 3. For example, in glaciated regions stratified drift comprising sandy material represents an unfavorable geochemical environment, unless topography is steep and hydromorphic dispersion is able to generate metal-rich zones. In contrast, nonstratified drift of variable composition presents less of a problem to the formation of geochemical anomalies, although interpretation of resulting geochemical patterns requires a knowledge of glacial history and directions of ice movement.

pH–Eh/Element mobility

Element mobility is considered in relation to local pH and Eh conditions (Table 3.3). Geochemical patterns for elements that are mainly dispersed clastically (such as Au, Pb and Cr) are compared to patterns for more chemically mobile elements (such as Mo, Cu, Zn and U). Possible geochemical barriers, resulting from marked changes in pH changes or oxidation/reduction conditions related to breaks in slope, seepages, bogs, etc., are evaluated with respect to changing metal abundances, ratios and extractability. It may then be possible to discount some geochemical features as probably being accumulations of metals, derived from background sources, at geochemical barriers.

By reviewing the influence of each factor in turn, some "areas of interest" will be attributed to natural processes unrelated to significant mineralization. These are eliminated from further consideration as "false anomalies". To confirm their validity, results for the remaining "areas of interest" are checked for freedom from obvious sampling and analytical errors (the six factors depicted in the upper left of the Flow Chart and discussed in Chapter 4). Results still considered to merit followup should then be confirmed by reanalysis (10% of the funding for chemical analysis should be reserved for this purpose).

Reanalysis, which is particularly important for reconnaissance surveys where anomalous conditions can be represented by single samples, will normally remove spurious results or features introduced by the laboratory (Figure 4.16). However, elimination of systematic errors introduced during sample preparation will require, depending on their origin, either reprocessing of sample rejects or a return to the field and resampling.

In addition to eliminating "areas of interest", assessment of one or more of the factors on the right hand side of Flow Chart 2 may indicate that part of the survey area was unsuited to geochemistry. This is of fundamental importance in deciding to retain or release ground on the basis of the geochemical data. Alternatively, it may become apparent that a modified methodology is more likely to succeed. For example, if a review of overburden type indicates that stratified drift is prevalent, normal soil sampling procedures may be inappropriate. If, however, there is sufficient topographic relief, a program of sampling and analysis designed to detect hydromorphic anomalies might be a relatively inexpensive method of detecting concealed mineral occurrences (Figure 6.4).

"Areas of interest" that survive the assessment of Flow Chart 2 are divided into the following categories at the bottom of the chart:

(1) primary seepage features;
(2) secondary seepage features (derived from transported metal-rich overburden);
(3) residual, in situ features;
(4) transported features, down ice or downslope of their bedrock source; and
(5) other features, some related to geochemical barriers, others related to sampling variability, analytical variability, etc.

Categories 1 to 4 are recognized as features having a source in bedrock and are now upgraded to "geochemical anomalies". In Flow Chart 3 (Figure 6.9) these are related to the most appropriate geochemical model as a guide to the probable disposition of the bedrock source(s). In this case, the left hand side of Figure 6.9 summarizes some of the conceptual models for glaciated terrains based on Bradshaw (1975). Appropriate models for other regions are to be found in Kauranne (1976), Lovering and McCarthy (1978) and Butt and Smith (1980).

Recognition of element zoning patterns, reflecting either their primary distribution in bedrock or differential mobility, and geochemical gradients are important at this stage. For example, with the "well drained glacial till" model in the second box of Figure 6.9, anomalous conditions developed in response to glacial transport will exhibit a sharp boundary at the up-ice "beginning" of the anomaly train as shown in Figure 6.3. Concentrations in the down ice direction will decline due to dilution, and the anomaly may become broader with distance from source.

Flow Chart 3 has related anomalies to proposed model(s) of their origin. In Flow Chart 4 (Figure 6.10) the bona fide "geochemical anomalies," suggesting proximity to a mineral occurrence, are numbered (catalogued) and described. It is useful to record anomaly dimensions and contrast between the anomaly and adjacent background expressed as a ratio. Directional trends are described, as these might reflect structure, lithology, alteration zones, etcetera. Any relationships between geochemical and geophysical anomalies should be incorporated at this stage. Geochemical patterns recognized as related to known mines or mineral occurrences may aid assessment of the significance of other patterns that may be related to new mineral occurrences. Finally, to arrive at a priority rating for followup, anomaly descriptions are compared to the requirements imposed by the geological model controlling overall exploration philosophy, exploration strategy, land status and economics.

ANOMALY FOLLOWUP

Prior to followup, the selected anomalies are examined in consultation with the geologist and geophysicist. Proposed geochemical activites are specified with respect to sampling and analysis and estimated costs. These must be approved or modified to meet budgetary constraints. The geochemist must give a convincing assessment of the data and be realistic as to the best approach for followup.

Perusal of many conceptual geochemical models shows that zones of maximum metal concentrations in soils are often laterally displaced from the suboutcrop of their bedrock source. Trenching or drilling of the strongest portion of the anomaly in the expectation of finding an underlying mineral occurrence is therefore misguided. Under these circumstances anomaly followup commonly involves two phases: (a) additional, fill-in sampling to better define and confirm the anomaly; and, when this has been achieved, (b) an attempt to trace the anomaly to its source or "roots" as indicated by local conditions and application of the relevant conceptual model.

Geochemical activities during the second phase of followup typically involve:

(1) *Complementary sampling:* involving collection of additional types of material such as humus or lithogeochemical samples.
(2) *Complementary analysis:* size fraction or partial extraction studies to distinguish mechanical and hydromorphic anomalies.
(3) *Depth studies:* trenching, pitting or overburden drilling accompanied by profile sampling to obtain a three dimensional view of metal distribution and dispersion patterns as a guide to the source of the anomaly.

The last is particularly important and, depending on local conditions and the geochemical model, will often extend upslope and/or up ice into areas that may be characterized by background values in near surface soils.

Geological followup always accompanies geochemical followup. The distribution and nature of outcrops are mapped and prospecting is conducted to detect evidence of the sought-after mineral occurrence and/or alteration. Geological units and potential ore controls are identified. Conformity of the

TABLE 6.1—Common failings encountered in followup of geochemical soil surveys.

Problem	Solution
1. The anomaly is assumed to be valid. No checks are made.	Check assumptions by re-analysis and resampling.
2. Anomalies are poorly defined.	Use a more detailed sampling plan.
3. The anomaly is defined by too many samples.	Sample density shuld be selected by a synthesis of geological target, cost, orientation and experience.
4a. Trenching and/or drilling of the anomaly site did not lead to discovery.	Mechanical and hydromorphic dispersion of metal to the site must be considered. The source lies upslope or up ice of the anomaly.
4b. Trenching of a contoured anomaly did not lead to discovery.	As in (4a) or the point source nature of geochemical data must be recognized.
5. The anomaly is "explained away" when drilling is unsuccessful.	Reinterpretation of the data is necessary. Failure to locate a source for a bona fide anomaly means the source remains to be discovered.
6. A minor mineral occurrence explains the anomaly, even though this is trivial considering the size/grade of the initial anomaly.	A second source probably remains to be found.
7. Rock types or alteration are not right where exposed near the anomaly.	The importance of geological observations must be tempered by uncertainty regarding what is not exposed.
8. The anomaly does not fit the geological model.	Refinements of exploration philosophy might be warranted.
9. Geophysical features of economic importance are not identified under the anomaly.	Was the right geophysical method used? Must the source have a geophysical response? Consider mechanical and hydromorphic dispersion as in (4).
10. Geophysical targets near the geochemical anomaly were tested and found to be uninteresting.	Determine if geophysical and geochemical anomalies had to be related.
11. Geophysical targets were tested in prference to source areas suggested by the geochemical interpretation.	Exploration bias is unavoidable. An independent audit of exploration procedures may be needed.

local geology to the geological model is assessed, and favorable indications increase the interest that can be generated. It is to be expected that minor mineral showings will be found. These might explain the anomalous geochemical conditions, but common sense must prevail if a minor bedrock prospect or boulder occurrence is being suggested as the only source of a large anomaly. Minor showings found at this stage of exploration are therefore best considered within the overall potential economic geology of the area. Exploration cannot be considered complete until the origin of an anomaly is fully explained.

Geophysical surveys used in followup of geochemical anomalies are usually directed to specific objectives. For example, use of magnetic surveys as guides to geology or electromagnetic surveys to locate conductors representing sulfide concentrations, graphite, and/or alteration zones and/or faults. Geophysical followup will not normally involve surveying of the entire soil grid area.

Experience has shown that many geochemical failures occur during the followup because of false assumptions or gross misinterpretation of the origin of the anomaly. Too often the soil anomaly is assumed to directly overlie a hoped-for ore zone, and the effort required to locate the real source of the anomaly is bypassed in favor of drilling. Some of the most common mistakes made in geochemical followup are summarized in Table 6.1.

REFERENCES

Bradshaw, P.M.D. (editor) 1975. Conceptual models in exploration geochemistry—the Canadian Cordillera and Canadian Shield. Journal Geochemical Exploration, v. 4, p. 2–213.

Butt, C.R.M. and Smith, R.E. (editors) 1980. Conceptual models in exploration geochemistry—Australia. Journal Geochemical Exploration, v. 12, p. 89–365.

Fortescue, J.A.C. 1980. Environmental Geochemistry. Springer-Verlag, New York, 347 pp.

Kauranne, L.K. (editor) 1976. Conceptual models in Exploration Geochemistry—Norden 1975. Journal Geochemical Exploration, v. 5, p. 173–420.

Lovering, T.G. and McCarthy, J.H. Jr. (editors) 1978. Conceptual models in exploration geochemistry—the Basin and Range Province of the Western United States and Northern Mexico. Journal Geochemical Exploration, v. 9, p. 113–276.

Chapter 7

CASE HISTORY AND PROBLEM 1:

THE TONKIN SPRINGS GOLD MINING DISTRICT, NEVADA, U.S.A.

M. B. Mehrtens

Geochemical techniques played a major role in exploration of the Tonkin Springs district and ultimately led to discovery of economically significant bodies of gold mineralization. Using some of the information obtained during the exploration program, it is possible to review the geochemical environment, secondary dispersion processes and survey techniques used successfully in this part of Nevada.

The Tonkin Springs district is located in west-central Eureka County, Nevada, within the Simpson Park Range approximately 65 km northwest of the town of Eureka (Figure 7.1). Topography is typical of the Basin and Range structural province being characterized by long narrow valleys and north easterly trending mountain ranges with elevations varying between 1,700 and 3,100 m. Precipitation is in the order of 400 mm per year, the major portion of which occurs in the higher elevations during winter and spring. Soils are light brown to brown desert soils of residual origin in locations above the gravel-filled valleys and pediments. Vegetation consists of sagebrush and sparse grass in the valleys with juniper, pinyon and mountain mahogany in the higher country.

> From what you know of Nevada and its environment, can you make a statement as to the type(s) of secondary dispersion process that are operating here?
>
> What local features of the secondary environment are likely to cause problems in a soil survey?

The district is underlain by Lower Paleozoic sedimentary rocks and by Eocene–Oligocene rhyolite to andesite tuffs and flows (Merriam and Anderson, 1942) (Figure 7.2). The known economic mineralization at Tonkin occurs above the pediment at elevations of between 2,075 and 2,150 m and is localized within imbricate zones of the Roberts Mountain thrust (Roberts, 1966) intersected by high-angle faults. Ore is preferentially developed within silicified shaly carbonate rocks. Mineralizing fluids appear to have accessed the rocks via near-vertical faults and fractures while low-angled thrust faults and their associated breccias provided excellent lateral permeability. Anticlinal areas within the gently folded thrust system appear to have been the most favored sites for ore deposition particularly where the host rocks are overlain by relatively impermeable material such as clay-altered latite sills and rhyolitic ash flows.

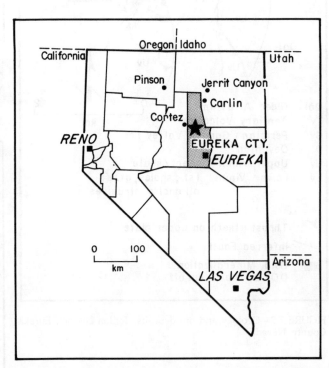

FIGURE 7.1—Regional location map, Tonkin Springs, Nevada.

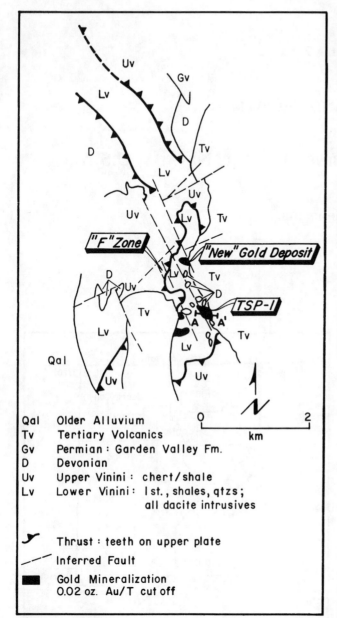

Qal	Older Alluvium
Tv	Tertiary Volcanics
Gv	Permian: Garden Valley Fm.
D	Devonian
Uv	Upper Vinini: chert/shale
Lv	Lower Vinini: lst., shales, qtzs; all dacite intrusives

FIGURE 7.2—Geology and ore deposits, Tonkin District, Eureka County, Nevada.

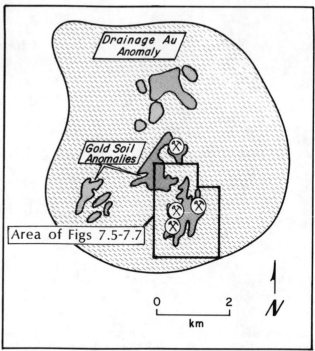

FIGURE 7.3—Distribution of gold in soils and drainage sediments within the Tonkin Springs District, Nevada.

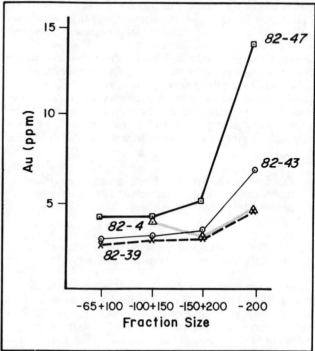

FIGURE 7.4—Gold distribution in oxide ore by fraction size.

Elevated concentrations of a number of elements related to the gold mineralization are detectable in rocks, soils and drainage sediments over several square kilometers around Tonkin Springs (Figure 7.3). This large surface geochemical expression is thought to reflect the scale of the hydrothermal (hot spring) system responsible for the precious metal mineralization of the district. It will be noted that the soil metal anomalies are enveloped by the drainage anomaly. Indeed it is this large drainage sediment dispersion pattern that allowed the initial appraisal of the Tonkin Springs district.

TABLE 7.1—Metal distribution in the soil profile and underlying bedrock in weakly mineralized and ore-bearing localities, Tonkin Springs, Nevada. Soil samples are routinely collected at 20–40 cm depth. All results in ppm.

Horizon	Depth (cm)	Weakly mineralized[1]						Ore bearing[2]			
		Au	Ag	Hg	As	Sb	Tl	Au	Hg	As	Sb
Residual soils											
Organic A	0–2	<0.02	<0.2	0.100	150	4	1.8				
Brown soil	2–20	0.03	<0.2	0.120	150	6	2.2				
Brown soil	20–40	0.05	<0.2	0.160	300	5	2.6	0.5	0.276	110	7
Brown soil	40–60	0.04	<0.2	0.115	500	7	3.3				
Bedrock											
	60–150	0.13						1.2		900	
	150–300	0.27						3.8		3000	
Average ore		4.5	3.4	3.6	900	28					

[1]Location 82-23 Figure 7.9
[2]Location 82-2 Figure 7.9

Stream sediment sampling led to recognition of the major mineralized areas, which were subsequently investigated by the soil survey described here.

In the unoxidized ore, gold occurs as submicron sized particles within framboidal pyrite accompanied by arsenic as orpiment and realgar and mercury in the form of cinnabar. Antimony occurs as small clusters of acicular stibnite. Oxidized ores contain free gold as finely divided particles with limonite and scorodite. Examination of the distribution of gold within various size fractions of crushed oxide ore

TABLE 7.2—Threshold values for Au, As, Sb and Hg in residual soils, 20–40 cm depth, Tonkin Springs, Nevada. Results in ppm.

Au	As	Sb	Hg
0.02	100	7	0.100

confirms the association of gold with the minus 200 mesh fraction (Figure 7.4).

The character of the oxidized ore has some clear implications for the type of sampling and sample preparation that might be applied in an exploration soil survey program. Try to list these and also explain how the host rocks and local secondary environment may cause you to modify your approach.

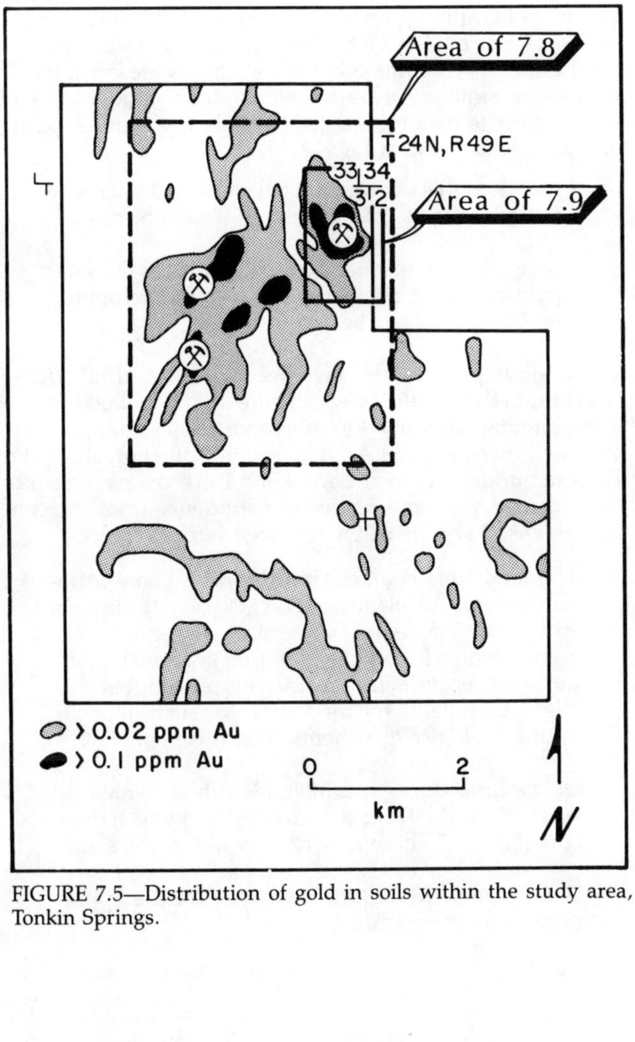

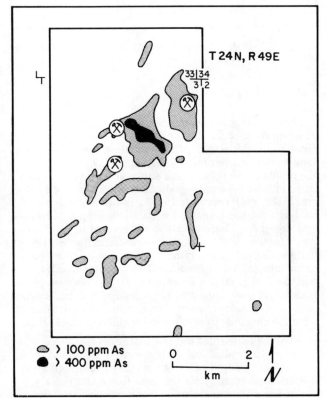

FIGURE 7.5—Distribution of gold in soils within the study area, Tonkin Springs.

FIGURE 7.6—Distribution of arsenic in soils within the study area, Tonkin Springs.

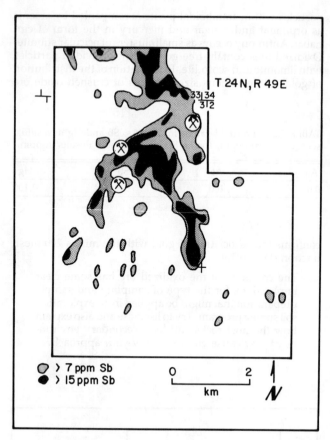

FIGURE 7.7—Distribution of antimony in soils within the study area, Tonkin Springs.

Above average concentrations of Au, Ag, Hg, As, Sb and Tl which characterize the Tonkin Springs mineralized zones, enter the residual soils upon oxidation and weathering. Limited data concerning the distribution of metals through the profile of the residual soil and near-surface bedrock are listed in Table 7.1 from weakly mineralized and ore-bearing areas. The results presented are for minus 80 mesh material analyzed for the contained metals following a hot mixed acid digestion. These data show a general but weak tendency for the metals to increase in concentration with depth in the soil profile and a rather abrupt increase from the soil to the mineralized near-surface bedrock. It is worth noting that the A organic-rich layer of the soil is impoverished in gold but generally indicative of mineralized bedrock, in terms of Hg and As values, when compared with threshold levels for these elements in routine soil samples (Table 7.2).

The data presented indicate that an acceptable geochemical contrast can be obtained by routine sampling at a depth of 20–40 cm. Can you design an orientation study for this area that could quickly confirm the suitability of this procedure and also reveal any opportunity for further optimizing survey procedures?

With the data available here can you suggest any supplementary or alternate procedures that would improve survey efficiency?

Detailed soil sampling at Tonkin Springs reveals the soil metal anomalies to be rather complex (Figures 7.5, 7.6, 7.7). One of these metal-rich locales occupies an area 1,000 × 2,000 m and consists of a number of composite Au, As, Sb, Hg anomalies (Figure 7.8). The high metal values tend to be aligned in two main directions; namely, NNW and NE, which closely approximate the two principal normal fault sets mapped in the bedrock. Except for Hg, composite metal soil anomalies are developed exclusively over limestones, shales and sandstones of the Lower Paleozoic. Mercury alone, however, is strongly enriched in soils over certain sections of the clay-altered Tertiary rhyolites and within a number of linear features over the Lower Paleozoic section.

Can you give any explanation for the unique behavior of mercury?

In Figure 7.9, attention is focused on a portion of the soil metal anomaly where economic grades of gold have been located. Examination of the gold data indicates that the 0.1 ppm and 1.00 ppm isopleths in soils and surface bedrock, respectively, define the bounds of near-surface gold ore fairly accurately. In contrast, deeper ore zones overlain by barren rocks, although within the limits of the soil anomaly, are not clearly evidenced by the gold values in surface soils. Furthermore, when the other ore elements are considered, results are similarly inconclusive. Indeed, gold values in soils appear to be a more reliable guide to suboutcropping ore than does soil data for Hg, Sb or As.

What does this information tell you about the mode of occurrence of gold in soils at Tonkin Springs?

There are a number of possible reasons why gold might not be a reliable guide to suboutcropping ore. Can you describe any of them?

Ore grade gold values have been intersected in drilling at a number of localities within the 1,000 × 2,000 m soil metal anomaly described in this report. These ore pods are indicated by coincident Au, Hg, As, Sb soil anomalies, and all occur under minimal cover. Blind ore deposits cannot be discerned with any degree of confidence from the conventional soil geochemistry reported here.

The large area characterized by raised concentrations of indicator elements (and gold) and the known structural complexity of the area suggest a very high potential for blind and buried gold ore bodies. Can you recommend any survey procedures that might be suitable for the search for such deposits in this geological/geochemical environment?

Of the procedures you have identified, which do you feel would be most successful and which most cost effective? These may or may not be the same.

What criteria would you use to select the appropriate survey procedure?

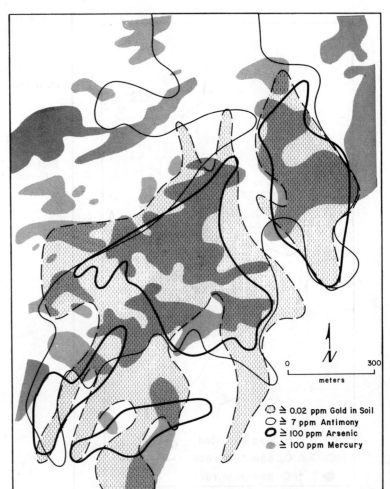

FIGURE 7.8—Distribution of gold, arsenic, antimony and mercury in soils within the detailed study area, Tonkin Springs.

This example of secondary dispersion of metals in a semi-arid region characterized by thin immature residual soils illustrates the close interrelationships metal anomalies in various surface media bear one to the other and the degree to which bedrock structure and mineralization are revealed by these data.

REFERENCES

Merriam, C.W. and Anderson, C.A. 1942. Reconnaissance survey of the Roberts Mountains, Nevada. Geological Society of America Bulletin, v. 53, p. 1675–1726.

Roberts, R.J. 1966. Metallogenic provinces and mineral belts in Nevada. Nevada Bureau of Mines Report 13A, p. 47–72.

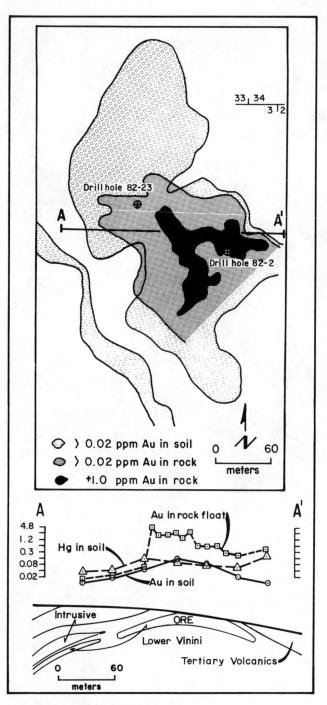

FIGURE 7.9—Distribution of gold in soils and surface bedrock over an ore-bearing locality, Tonkin Springs.

Chapter 8

CASE HISTORY AND PROBLEM 2:

COED–Y–BRENIN PORPHYRY COPPER, NORTH WALES, GREAT BRITAIN

M. B. Mehrtens

Geochemistry was used extensively during the exploration program by Riofinex that led to discovery of the Coed–y–Brenin porphyry copper deposit. The attention of mineral explorers had been drawn to the area because of a small bog known as the Turf Copper Works, which in the 1860's produced copper from peat-ash shipped to the Swansea refineries. The bog occurs in the center of the district and has been a starting point for most investigators, of which Riofinex was among the most recent and successful.

Coed–y–Brenin is located 7.25 km north of Dolgellau on the southeast flank of the Harlech dome in the County of Gwynedd (Figure 8.1). The country undulates at elevations of between 175 and 250 meters O.D. and is drained by the Afon Wen and Afon Mawddach, which occupy deeply incised valleys and flow to the south and southwest to the Mawddach estuary. The region is underlain by upper Cambrian turbidites and argillites that have been intruded by diorite sills, dikes and stocks that are thought to be co-magmatic with the nearby Rhobell Fawr calc-alkaline volcanics of lower Ordovician age (Figure 8.2). Rock exposure is severely limited with approximately 95% of all bedrock concealed beneath Pleistocene glacial till that has an average thickness of about 6 meters with local accumulations greater than 30 meters. The till was deposited by ice that moved from the north to

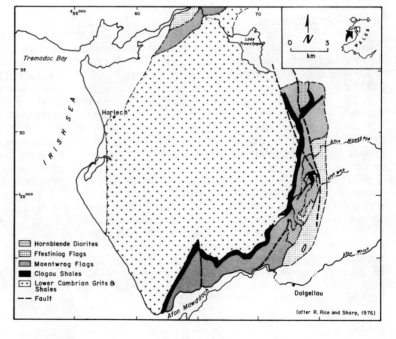

FIGURE 8.1—Geological sketch map of the area north of Dolgellau, Wales. After Rice and Sharp (1976).

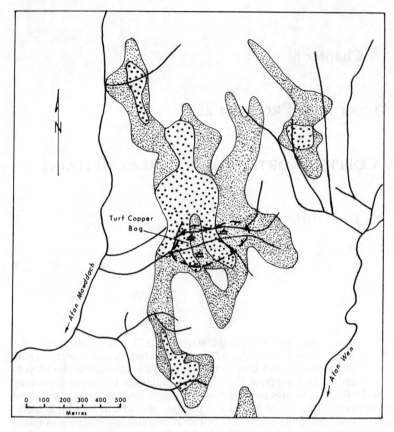

FIGURE 8.2—Coed–y–Brenin Copper Deposit and associated Cu–Mo soil anomalies, North Wales. Areas with anomalous Mo values are also anomalous for Cu.

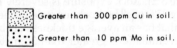

south. Soils are predominantly brown podzols. The indigenous flora include hazel, oak and other deciduous trees. These have largely been supplanted by coniferous trees as a consequence of a reafforestation program. The climate is temperate with annual precipitation in the range 1,500–2,000 mm.

A 150 × 60 m survey grid was established across the property and a systematic soil geochemical survey carried out. The B horizon of soils was sampled at a depth of between 10 and 20 cm and the minus 80 mesh fraction analyzed for copper and molybdenum using a strong hot acid digestion technique.

Soil copper and molybdenum anomalies (Figure 8.2) surround the Turf Copper Bog, extending in a roughly linear pattern oriented north-south parallel with the direction of ice transport. The soil anomaly is underlain, at least in part, by glacial till rich in mineralized fragments (Figure 8.3). It was on the basis of the last ice transport direction that the bedrock source for the metals was initially sought just to the north (the up-ice termination) of the macro ore-boulder train. Drilling in this area was, however, unsuccessful in locating mineralization in bedrock.

There were at least two mistakes; first, the soil geochemistry is based entirely on hot extractable metal data when the ratio of cold to hot extractable metal values would have been useful to obtain a better understanding of the soil anomaly. That is to say, these additional data may have allowed an estimate of the extent to which the metal in the overburden was incorporated by mechanical (ice) transport as opposed to the metal added to the overburden by hydromorphic processes. Secondly, and more important, insufficient weight was given (during the initial drilling) to the evidence that copper was being dispersed in mineralized groundwater to be deposited in bogs and at seepage sites peripheral to a concealed deposit (Figure 8.3).

What evidence can you find in the information presented here that copper is being dispensed in ground water?

What additional information not presented here, but readily available, would substantially improve your ability to interpret the soils data and might aid in recognition of hydromorphic dispersion patterns?

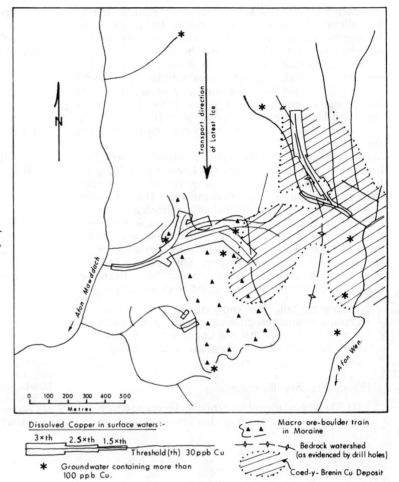

FIGURE 8.3—Macro ore-boulder train and Cu anomalies in ground and surface waters, Coed–y–Brenin, North Wales.

In the light of this recognition, groundwater was sampled at seepages on the property as well as surface waters from streams draining the area of interest. These were analyzed in the field for dissolved copper yielding the results shown in Figure 8.3.

> Do you know how the analysis for dissolved Cu in water is performed and what additional analytical parameter, easily determined in the field, would be useful at this time?

The data show the presence of raised copper values well to the east of the soil anomaly. In fact the geochemical data now indicate dispersion away from a major source of copper mineralization lying buried to the east of the turf bog.

In this connection it should be added that the Riofinex drilling showed that the Turf Copper valley was originally occupied by the "pre-glacial Afon Wen". As one consequence of the last glacial episode, the old gravel-filled stream bed was overlain by glacial till and the Afon Wen rerouted in its present course following retreat of the ice (Figure 8.4).

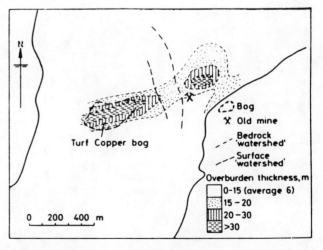

FIGURE 8.4—Thickness of overburden in the vicinity of the "Turf Copper Bog", North Wales. After Mehrtens et al. (1973).

The buried gravel-filled valley thus provides an additional but hitherto unsuspected passageway for groundwater leaving the oxidizing deposit to drain to the west in an area where surface water flows toward the east.

The soil anomaly, therefore, is a complex product of mechanically (ice) incorporated metal sulfides together with Cu and Mo added by hydromorphic processes from oxidizing bedrock mineralization lying to the east and largely concealed beneath barren till. The bedrock source of the mineralized fragments mechanically dispersed in the till is still not known.

The copper-in-water (Cu aq) data (Figure 8.3) obtained in the field at seepage sites are the key to determining the location of the concealed mineral deposit ultimately encountered in the Riofinex drill program. These data also imply a second but unexplored mineralized area not far north of the drilled deposit. It seems reasonable to suggest that a portion of the metal in the soil anomaly and most of the mineralized fragments in the till were derived from this site as a result of glacial transport. Environmental restrictions, however, caused Riofinex to abandon the project before this northern prospective area could be investigated.

> From what you now know of the Riofinex program, can you recommend specific geochemical survey procedures that will improve the effectiveness of any further exploration in this environment?

This case history illustrates:

(1) The importance of recognizing, before a survey begins, the nature of dispersion processes operating in an area so that sampling and analytical techniques may be chosen that will elucidate these processes.
(2) That complicated and potentially confusing secondary dispersion patterns may develop from mechanical and hydromorphic dispersion of metals and the importance of identifying processes and pathways (plumbing systems) when interpreting these patterns.
(3) The need to consider all aspects of the landscape and draw upon all available information when interpreting geochemical data.
(4) That it may be necessary to use a variety of techniques (sample medium, analytical procedures, etc.) in order to locate the source of anomalous metal in environments with complex overburden conditions.
(5) That a source of anomalous metal, when found must fully explain the geochemical features under investigation before exploration can be considered complete.

REFERENCES

Mehrtens, M.B., Tooms, J.S. and Troup, A.G. 1973. Geochemical dispersion from base metal mineralization. In: Jones, M.J. (editor), Geochemical Exploration 1972. Institute of Mining and Metallurgy, London, p. 105–115.

Rice, R. and Sharp, G.J. 1976. Copper mineralization in the forest of Coed–y–Brenin, North Wales. Transactions Institute Mining and Metallurgy Section B, v. 85, p. B1–B13.

Chapter 9

CASE HISTORY AND PROBLEM 3:

THE VOLCANOGENIC MASSIVE-SULFIDE TARGET

S. J. Hoffman

PRELIMINARY STUDIES

The exploration target is a Noranda-type volcanogenic massive-sulfide (VMS) deposit. Property history illustrates one classical exploration approach in greenstone belts of the Canadian Shield. An airborne electromagnetic (EM) survey was conducted and a series of conductors were defined. These were followed up on the ground by establishing a cut grid using a 100 m line spacing and a 25 m picket interval. Conductor axes, located by horizontal loop EM, would then normally be drill tested at several locations if the general geology was favorable. Unfortunately, the majority of conductors are usually found to be graphite or barren pyrite/pyrrhotite horizons.

The geological model of the VMS deposit type is illustrated in Figure 9.1. The deposit typically consists of massive accumulations of pyrite, pyrrhotite, sphalerite, galena, chalcopyrite and barite with lesser amounts of Ag and Au. Very high Hg contents are often associated with the sphalerite. The surrounding volcanic host rocks, particularly in the footwall, are usually enriched in Mg and depleted in Na and Ca (Figure 9.2). Geochemical behavior of these elements in the surficial environment is summarized in Table 3.3.

1. Prepare a table, such as that for unconformity U deposits in Table 2.4, to describe conditions you consider necessary to define a drill target for VMS deposits.

(Answers to questions appear at the end of the chapter).

FIELD ORIENTATION

The landscape is relatively flat with elevation differences of about 25 m per kilometre. These differences increase in the east to 50 m over 500 m as a major drainage system is

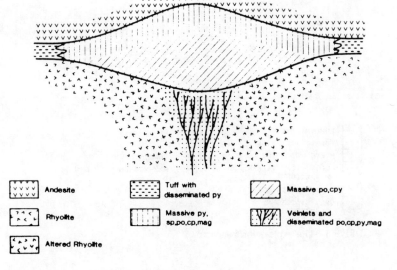

FIGURE 9.1—Idealized section through an Archean massive sulfide deposit of the Abitibi belt after Franklin et al. (1981). Reproduced from Economic Geology, 1981, Seventy-fifth Anniversary Volume 1903–1980, p. 570.

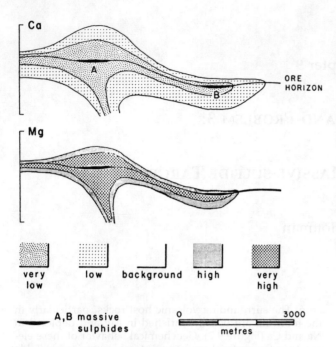

FIGURE 9.2—Schematic distribution of Ca and Mg in felsic volcanic rocks around proximal (A) and distal (B) massive sulfide deposits, New Brunswick. From Govett and Nichol (1979). Reproduced with permission of the Minister of Supply and Services, Canada.

approached. There is deciduous forest cover, and bedrock is poorly exposed with outcrops representing about 1% of the landscape. Traversing along roads crossing the property indicated that overburden was essentially residual and 1 or 2 m thick. Soils exhibit standard profile development with a horizon sequence LH–AE–BF–BM–C1 in which the BF horizon, preferred for geochemical sampling, is at a depth of 20 to 30 cm. Coarse fragments in the soil consist of angular, 2 cm fragments of a single rock type. Examination of fragments would certainly assist in geological mapping.

Based on field observations, a soil geochemical survey was proposed to evaluate the major geophysical anomaly on the property. The objective was to define metal-rich portions of the conductor, which could then serve to focus the diamond drill program. The sample interval selected was 25 m within 75 m of the axis of the conductor, increasing to 50 m further away. The BF soil horizon was chosen for sampling in the absence of orientation studies.

2. Did the selection of sample interval along each line optimize the sampling program? Explain your answer with reference to the standard overburden geochemical models.

3. Does the occurrence of a geochemical anomaly associated with the conductor play a role in your model for selecting drill targets (see your response to Question 1)?

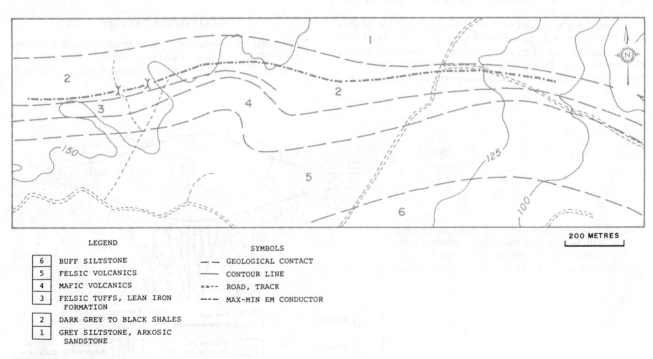

FIGURE 9.3—The exploration basemap.

CONTINUED OFFICE PLANNING

Basemaps showing topography, geology, conductor axis, roads and stream channelways were prepared (Figure 9.3). Samplers were relatively well trained and experienced in soil sampling procedures.

PROPERTY EVALUATION

The soil survey was completed according to plan. Samples were analyzed, following an aqua regia digestion, for 30 elements using an ICP determination method. Results for 18 elements exhibiting significant variations are plotted in Figure 9.4 A–R.

4. Are there any sampling artifacts in the data?

5. Do you see any analytical problems with the data set?

6. Can you recognize any geochemical patterns that might be indicative of underlying geological controls?

7. Based on Figure 9.4, where would you drill test the conductor? Assume that a VMS deposit 0.5 km long would be economically interesting. If you decide not to drill, what followup program would you propose?

8. Was a soil survey appropriate for this property? Refer to your response to Question 7 in answering this question.

ANOMALY FOLLOWUP—DRILL TESTING

The geochemical survey did not provide outstanding geochemical anomalies for Cu, Pb or Zn. An electromagnetic survey indicated that depth to the top of the conductor was likely to be less than 3 m. In view of the flat topography, thin overburden, easy access and low costs, recovery of bedrock samples from the conductor was considered feasible with a backhoe. Bedrock was intersected at two of three locations, shown in Figure 9.3 and 9.4, with the conductor being identified in one trench. At the third location, trenching was abandoned at 4 m. Overburden resembled that seen in soil pits and roadcuts.

9. Once trenching had been authorized, what procedures would you use to maximize collection of information in the event the project was to continue?

10. What geochemical barriers or other controls can you suggest that could have prevented dispersion and/or accumulation of Cu, Pb and Zn from bedrock into overlying soils under conditions described for this property?

11. Reviewing soil trace element distributions, a number of isolated, single point anomalies were defined throughout the grid for most elements. Can you explain the significance of these results in view of work described thus far?

12. The only observation made by the actual exploration program in relation to Question 9 was to note graphitic shales at the bottom of the trench intersecting a conductor. Basal overburden samples were not taken. Would you proceed to diamond drilling?

The exploration program did continue with diamond drilling.

13. What do you think the diamond drill program found (geologically and geochemically)?

ANSWERS

1. See Table 9.1.

2. No. This sampling plan will only work if topography is flat, and a residual anomaly has developed immediately over the ore zone (i.e., assuming no lateral displacement downslope or down ice). Both factors would have to be considered on a line-by-line basis to optimize the extra effort of the detailed sampling. The exploration program probably would not suffer from its omission.

3. Yes. If a significant soil anomaly can be defined near the conductor, as indicated by Case 3 of Table 9.1, and interpretation suggests bedrock at or near the conductor is the probable source, a drill target would be established. Interpretation would be facilitated by thin overburden, but this is not essential. For example, seepage anomalies some distance downslope of the conductor might be sufficient evidence of a potential massive-sulfide deposit to merit upgrading of a conductor to drill target status.

TABLE 9.1—Question 1: combinations of parameters required to establish a drill target for volcanogenic massive-sulfide deposits.

Parameter	Possible drill targets			
	Case 1	Case 2	Case 3	Case 4
Favorable geology	X	X	X	X
Conductor	X			X
Sulfide prospect		X		
Favorable alteration		X		
Base metal anomaly—bedrock		X		
Base metal anomaly—overburden			X	X
Mercury anomaly				
Thin overburden			X	X

X = essential criteria

4. Yes. The Mn distribution exhibits many high frequency values suggesting highly variable sample composition in the BF horizon. Elements such as Cu, Zn and As show some sporadically enhanced values, but there are too few to be considered a problem.

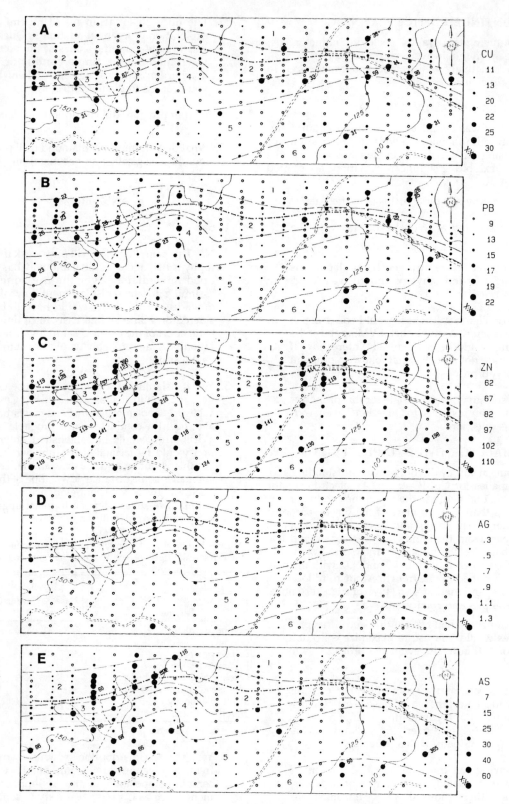

FIGURE 9.4 A–R—Geochemical results for 18 elements. Determinations by inductively coupled plasma spectroscopy after aqua regia decomposition.

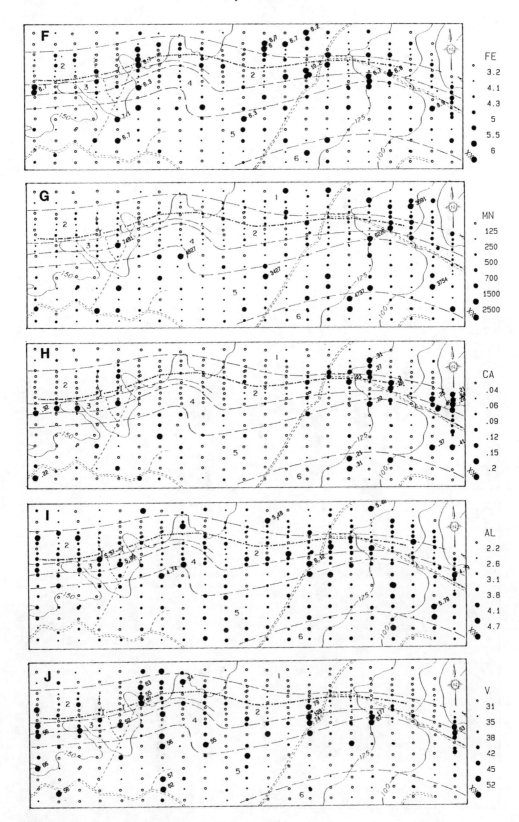

FIGURE 9.4—Geochemical results for 18 elements (continued).

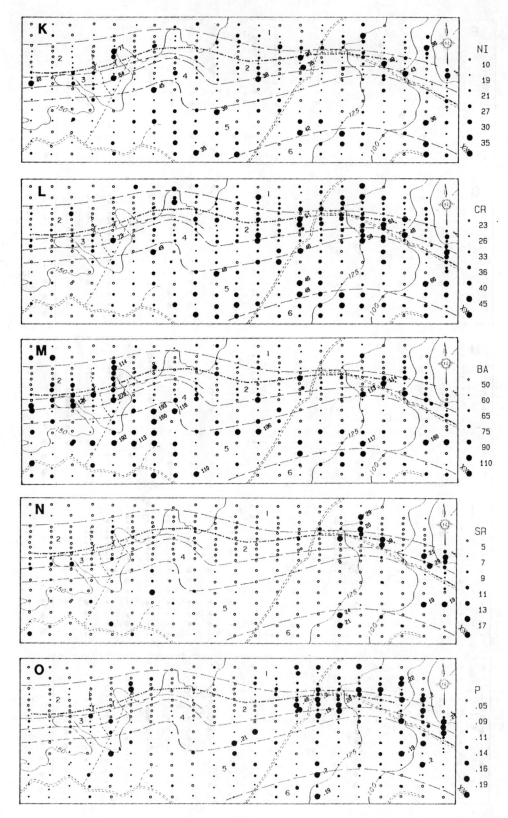

FIGURE 9.4—Geochemical results for 18 elements (continued).

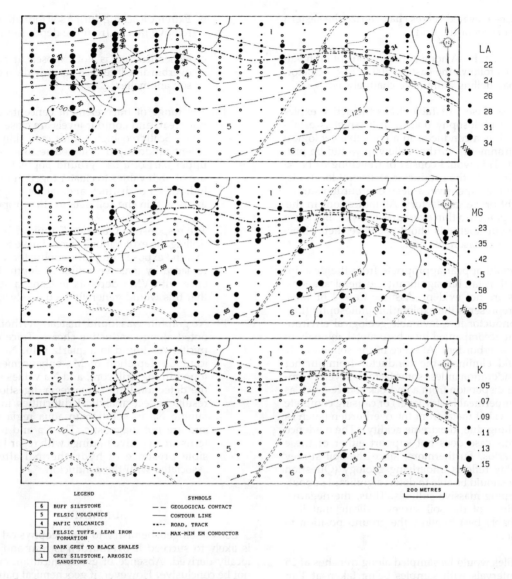

FIGURE 9.4—Geochemical results for 18 elements (continued).

5. No. Systematic enhancement is noted for As along Line 6 and V backgrounds are high on line 7, but serious systematic analytical errors are absent.

6. Geochemical units can be defined as features exhibiting either a general enhancement or depletion in metal contents over a broad area. Anomalies by contrast would be smaller features exhibiting a greater degree of contrast with local background and having a more jagged appearance when plotted as profiles. The following patterns might reflect geological controls:

a. As, La—enhanced over the western portion of the grid crosscutting geological trends, La is also depleted in the southeast.

b. Mn—enhanced in the northeast corner of the grid. Also heterogeneously enhanced in association with unit 5 felsic volcanics and soils immediately downslope of this unit.

c. V—enhanced overlying unit 4 mafic volcanics and unit 3 felsic volcanics and lean iron formation. Unit 5 and unit 1 gray siltstone and arkosic sandstones are relatively impoverished in V.

d. Cr—elevated values characterize the eastern portion of the grid.

e. Sr, Ca, P—A homogeneous zone of high values follows the eastern 700 m of conductor, parallelling a drainage channelway.

7. Unless Case 1 of Table 9.1 is the active exploration philosophy (i.e., a drill target is established by a conductor in a favorable geological environment), further followup would be prudent. This would consist of trenching or backhoeing into the conductor at several locations, or, if too expensive, a deep overburden survey might prove satisfactory. All indications in this case point to a thin overburden cover, and these followup methods should achieve their objective at a reasonable cost.

8. It would appear appropriate from the description thus far, but significant geochemical anomalies are not evident. Assume Case 4 of Table 9.1 represents the exploration philosophy (i.e., a conductor in a favorable geological environment accompanied by a base metal anomaly in thin overburden). Absence of the soil anomaly would eliminate the conductor from further consideration unless some factor is severely limiting dispersion from a bedrock source (such as an impermeable clay layer) or causing complete removal of metal from the overburden (such as leaching due to extreme acidity). The evidence in this case does not support either of these scenarios. Furthermore, since the soil survey weakly reflects underlying geology (Question 6), it should be able to detect significant suboutcropping massive sulfides. Thus, the negative findings of the soil survey indicate that it is probably best to allow the ground position to lapse.

9. Profiles would be sampled along trenches at 25 m intervals with samples being taken at 1 m intervals down profiles. This is designed to look for root zones of surface geochemical anomalies. If leaching of metals at surface is complete or overburden conditions do not favor geochemical dispersion from a bedrock source into the soil, the profile sampling, particularly the basal sample above bedrock, should provide an indication of a mineral occurrence nearby. Continuous chip samples would also be taken of trenched bedrock.

10. a. Active oxidation of pyrite from a massive sulfide might generate sufficient acid to mobilize Cu and Zn out of the soil and into the groundwater to be transported perhaps beyond the limits of the grid.

b. Extreme leaching during soil formation might remove metals from the depth sampled.

c. Mineralization is not exposed at bedrock/overburden interface to provide metal for soil anomalies.

11. The range of trace element concentrations is not great, and contour levels are close to each other. Scattered high values may simply represent randomly distributed results from the upper part of a background population. Failure of high values to cluster together probably indicates no distinct anomalous population exists within the soil data for the property.

12. The response to Question 7 applies. To justify drilling the explorationist would have to believe that the exposed graphitic conductor might be replaced by massive sulfide at depth. Experience with VMS deposits in Canada suggests that this would generally be unlikely.

13. No graphite was found; instead pyrrhotite was noted along conductive shears. Trace chalcopyrite was found in a quartz vein. No significant Ca depletion or Mg enhancement was found in the drill core, and Na values were at the detection limit (aqua regia digestion). Na results reflect analytical difficulties in maintaining Na in solution in aqua regia digests and are not diagnostic. The Ca and Mg patterns are negative findings with regard to possible presence of hydrothermal alteration associated with the conductor.

SUMMARY

Use of soil geochemistry to rate geophysical conductors is likely to succeed if overburden is thin and residual or locally derived. Absence of geochemical anomalies might not be conclusive. However, if geochemical patterns reflect the underlying geology, as they do in this case, they should also be capable of indicating presence of a significant suboutcropping mineral occurrence providing that sample density has been optimized and factors (such as leaching associated with extreme acidity) suppressing anomalous conditions have been shown to be absent.

REFERENCES

Franklin, J.M., Lyndon, J.W. and Sangster, D.F. 1981. Volcanic associated massive sulphide deposits. In: Skinner, B. J. (editor), Economic Geology Seventy-fifth Anniversary Volume 1905–1980, p. 485–627.

Govett, G.J.S. and Nichol, I. 1979. Lithogeochemistry. In: Hood, P. J. (editor), Geophysics and Geochemistry in the Search for Metallic Ores. Geological Survey of Canada, Economic Geology Report 31, p. 337–362.

Chapter 10

CASE HISTORY AND PROBLEM 4:

THE VOLCANOGENIC MASSIVE SULFIDE, A SECOND EXAMPLE

S. J. Hoffman

PRELIMINARY STUDIES

The target is another volcanogenic massive-sulfide (VMS) deposit. Public domain documents indicated many conductors on the property, which lies in a camp of past producing VMS deposits.

FIELD OBSERVATIONS

Landscape, overburden and soil conditions are similar to those described in Chapter 9. An existing grid, perpendicular to known conductors, was used to control soil geochemical work. Line spacing is approximately 120 m (400 feet), and pickets are in place every 15 m (50 feet). Road access is not as good as in the previous example. Geology and drainage are summarized in Figure 10.1: maximum relief is about 40 m.

CONTINUED OFFICE PLANNING

At this time the detailed location of conductor axes had not been established on the ground. A 60 m (200 foot) sample interval was chosen to evaluate property geochemistry.

PROPERTY EVALUATION

Soil samples were analyzed for 30 elements. Approximately one sixth of the survey area is reproduced for 21 elements and 11 field parameters in Figure 10.2. Metal-rich zones contoured in Figure 10.2 were defined by highlighting the upper 10% of values for each element from the entire survey area. As a result some maps, such as that for Ag (Figure 10.2D), have no multisample features whereas others, such as Fe (Figure 10.2N), exhibit large zones of geochemical enhancement.

1. Can you identify any geochemical anomalies suggestive of proximity to a VMS deposit? (Answers to questions appear at the end of the chapter).

2. Are there any geochemical distributions indicative of the underlying geology.

3. Are there any sampling artifacts in the data? Mark suspect samples with a cross through the sample point.

4. Do you see any analytical problems with the data set?

5. Referring to the geological model, are there any exploration targets on the property?

6. A massive sulfide occurrence suboutcrops within 500 m of this grid area. Does this change your answer to question 5? Why? Outline potential target areas.

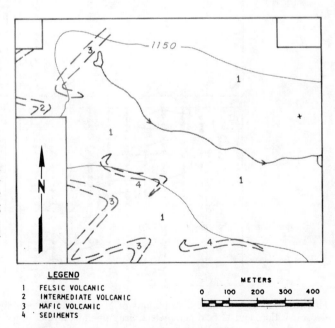

LEGEND
1 FELSIC VOLCANIC
2 INTERMEDIATE VOLCANIC
3 MAFIC VOLCANIC
4 SEDIMENTS

FIGURE 10.1—Property geology summarized on a topographic basemap.

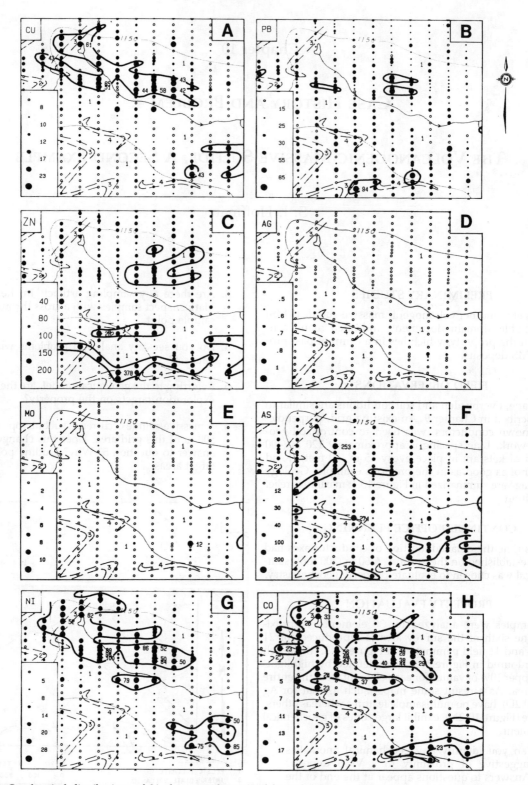

FIGURE 10.2—Geochemical distributions of 21 elements, determined by inductively coupled plasma spectroscopy after aqua regia decomposition, and 11 field parameters. See Appendix II of Chapter 3 for explanation of field parameter codes.

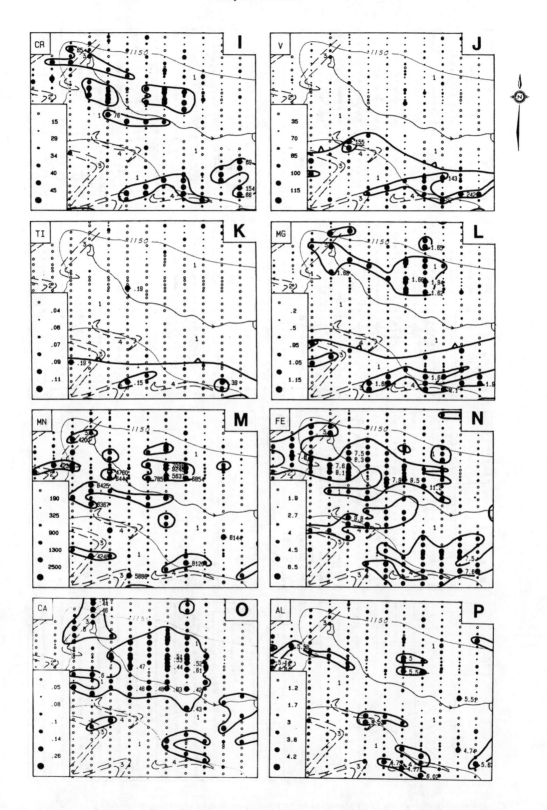

FIGURE 10.2—Geochemical distributions of 21 elements and 11 field parameters (continued).

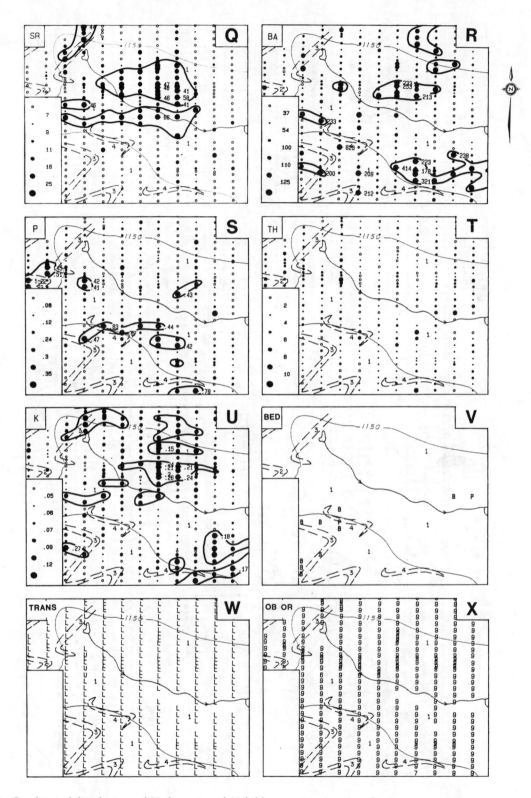

FIGURE 10.2—Geochemical distributions of 21 elements and 11 field parameters (continued).

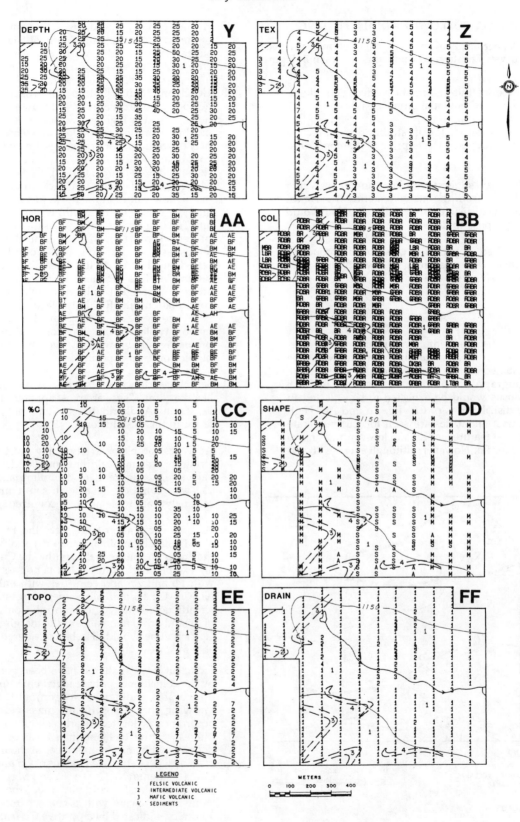

FIGURE 10.2—Geochemical distributions of 21 elements and 11 field parameters (continued).

ANOMALY FOLLOWUP—DRILL TESTING

EM results became available after completion of the geochemical survey. Location of conductors is shown in Figure 10.3.

7. How does this information affect your determination of potential target areas?

8. What, if any, followup studies do you recommend on these targets prior to diamond drilling?

9. Would you suggest examining and/or analyzing drill core from the massive sulfide occurrence if it was available to you? If yes, what would you propose to do with the data?

10. Are there any drill targets on this property? Explain your reasoning and what you would be looking for.

ANSWERS

1. In the absence of information on mineral occurrences in the area, geochemical data would be evaluated on the basis of geology after taking topography into account. Presence of VMS deposits would be expected to be reflected in the distributions of Cu, Pb, Zn and Ag. High contrast anomalies are seen for Cu, moderate contrast features for Zn and low contrast anomalies for Pb. However, although average Pb levels are regionally anomalous, at first glance, none of the anomalies are sufficiently outstanding to suggest they are indicating an extensive suboutcropping of massive sulfide.

2. Geochemical distribution patterns reflecting geological controls are typically areas of broad, homogeneously high or low metal values. Based on this definition the following elements display distribution patterns suggesting geological control in the north-central and southeast portions of the map area: Cu, Ni, Co, Cr, Fe and Mg. A marked change in backgrounds, from mainly low values in the north to a mix of high and low values in the south, is also seen for V and Ti. High background values might be interpreted as representing areas underlain by andesites, whereas low backgrounds might reflect rhyolitic or dacitic volcanics or derived sediments. This geochemical interpretation, in an area of residual soils, has some geological confirmation but, for the most part, provides new findings augmenting the limited information available from the geological map and providing possible working hypotheses in areas of poor outcrop exposure.

3. Sampling artifacts are suggested by "measles-like" patterns. These are most apparent for Mn, Fe, Ca and Al distributions but are not a major factor in this case history.

4. Analytical problems are typically indicated by strings of systematically enhanced or depressed values (e.g., Figure 4.16). Sampling and analysis were conducted in a north-south direction. No analytical problems are recognized.

5. Possible exploration targets are the Pb anomalies (residual, minimal transport) and perhaps the linear Zn feature in the south. The Cu anomaly exhibits high contrast but is not associated with Pb and Zn and coincides with enhanced Ni, Co, Cr, Fe and Mg values suggesting, as already noted, lithological control. Coincidence of Pb and/or Zn anomalies at the margins of the Cu-rich zone would also be considered favorable for prospecting.

6. Sufficient interest should have been generated by the coincidence of base metal enrichment in a favorable geological environment. If not, knowing that a significant massive sulfide occurrence is located nearby should increase interest. Figure 10.4A outlines areas of interest defined by Pb and Zn anomalies at the margins of areas where enhanced values of Cu and associated elements (Ni, Co, Cr, Fe and Mg) suggest presence of mafic volcanics (i.e., the Pb–Zn anomalies coincide with a change from mafic volcanics, indicated by a positive Cu feature, to felsic volcanics with relatively low Cu values).

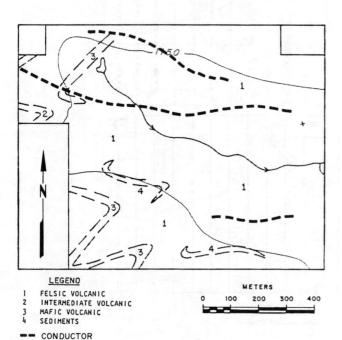

FIGURE 10.3—Location of EM conductors on the property.

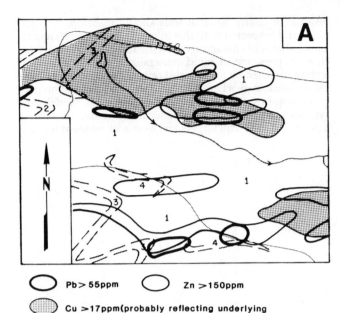

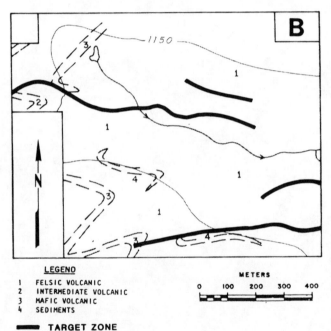

FIGURE 10.4—A: Summary of geochemical anomalies for Cu, Pb and Zn. B: Location of exploration targets.

Figure 10.4B outlines the linear exploration targets formed by this zone.

7. Excellent correspondence is seen between the location of conductors and geochemical target zones summarized in Figure 10.4B. Based on distribution of Cu, conductors without an anomalous Pb–Zn signature may only be reflecting barren contacts between mafic and felsic lithologies. Conductors are thus priority rated, the coincidence of anomalous geochemical and geophysical features suggesting potential drill targets.

8. Trenching (Figure 10.5) or deep overburden surveys are recommended to locate the root zones of the soil anomaly and possibly intersect a suboutcropping massive sulfide occurrence. This would be relatively inexpensive, and results should greatly assist in planning drill hole locations.

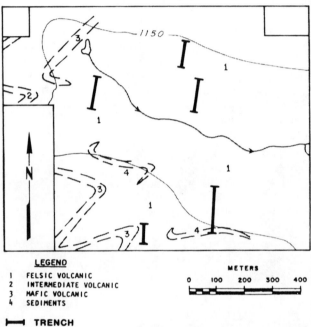

FIGURE 10.5—Targets for trenching based on a synthesis of geochemical anomalies from Figure 10.4B and EM conductors of Figure 10.3. Drill sites would be in the general vicinity of the proposed trenches if results were encouraging.

9. Yes. Multielement determinations (probably using ICP on an aqua regia or similar leach) could be used to identify distinctive alteration halos or base metal and pathfinder signatures around known VMS intersections. Depending on element mobility similar relationships might be anticipated in residual or locally derived glacial overburden. Such patterns would further refine target definition.

10. The decision to diamond drill would require a favorable outcome to the trenching program (Question 8): five targets can only be tentatively identified at this stage.

SUMMARY

The number of EM anomalies makes diamond drill testing rather expensive, and some method of rating them is desirable. The geological model suggests that many VMS deposits lie along or near the contact between felsic and mafic volcanics. Location of a deposit within this zone may be further defined by halos and distribution patterns for ore, alteration and pathfinder elements. Soil surveys in areas of residual or nearly residual soils should therefore be organized and inspected with this in mind. Low contrast anomalies or halos of characteristic elements may provide useful clues to presence of blind mineralization. Diamond drill testing can then be used to test only those conductors having a favorable geochemical signature. The exploration program requires confidence in the geological model and synthesis of geological, geochemical and geophysical information to define drill targets and assess subsequent results.

Chapter 11

CASE HISTORY AND PROBLEM 5: A COPPER PROPERTY

S. J. Hoffman

PRELIMINARY STUDIES

In this example ground was acquired as a result of regional geochemical exploration for alkaline Cu–Au porphyry deposits hosted in volcanic rocks. However, this geological model and its geochemical implications are irrelevant because followup of geochemical anomalies led to discovery of a totally different target type.

The property is located in the mountains of central British Columbia and extends from a forested valley floor to alpine meadows and bare rock on ridges and summits. Access is by helicopter to the alpine sections: a road runs in the valley 1 km south of the southern boundary of the property. Geological, geophysical and topographic information available at the time of discovery are given in Figure 11.1. The study dates from 1973 to 1978, before the advent of routine multielement analysis with the ICP, and analytical results are available only for Mo, Cu and Zn because costs for additional elements precluded their determination on a routine basis. Figure 11.2 displays a portion of the original regional stream sediment survey.

Comparison of Figures 11.1 and 11.2 identifies a stream sediment Cu and Mo anomaly apparently associated with an ultramafic body. Ground inspection confirmed the underlying geology to be a variety of ultramafic lithologies, including dunite, pyroxenite, peridotite, hornblendite and gabbro. It was therefore assumed that the Cu anomaly could be explained by high background Cu contents in an ultramafic unit, not an unusual finding (see tables for Cu in Levinson (1980) or Rose, Hawkes and Webb (1980)). This also provided an explanation of why the property had no prior history of prospecting or exploration by others. However, the Mo anomaly could not be accounted for by this geological scenario.

Followup was continued to locate a source for the Mo. Diorite float containing chalcopyrite and molybdenite was found. Once the first sulfide-rich boulder was located, many more were uncovered in the high alpine region. The stage was set for a major program of exploration.

FIELD ORIENTATION

Field orientation studies and other observations were made concurrently with anomaly followup (Figure 11.3). The property covers a series of mountain ridges where outcrop is relatively continuously exposed. These are flanked by prominent talus aprons. Glacial erosion and deposition occurred in mountain cirques and along major valleys, which now contain variable thicknesses of till.

North Cirque

The North Cirque trends northwest. Overburden is dominated by talus and landslide debris from the predominantly ultramafic hills in the east and andesitic hills in the south and west. Locally, concentrations of acidic intrusive boulders can contain abundant chalcopyrite (1–2% Cu). Overburden cover is probably less than 3 m thick at the cirque headwall and outcrop is intermittently exposed. Solifluction and seepage zones are common near the headwall in the southeast, and a prominent solifluction lobe follows the middle of the cirque valley. Overburden thickness exceeds 30 m in the banks of creeks cut into the solifluction lobe.

North Creek

The steep slopes carry talus deposits with a high proportion of fine material between angular ultramafic blocks. Outcrop is intermittently exposed, particularly along creek cuts. No vegetation grows on these slopes.

North Tip

The change in geology from ultramafics to andesites and granites is accompanied by the growth of vegetation that stabilizes the overburden. A small cirque incised into ultramafic bedrock on the east side of the ridge is cut by numerous dikes of intrusive rock, some containing chalcopyrite.

Tabletop Highlands
(2 km east of North Cirque on Figure 11.3)

This is a large area of rolling landscape, devoid of vegetation, with residual overburden consisting of blocks and fines of ultramafic material. Locally there are dikes of acidic composition.

South Cirque

South Cirque trends southeast and is physiographically similar to North Cirque but lacks a solifluction lobe. Mineralized intrusive blocks containing chalcopyrite and molybdenite are locally common. Outcrop is present intermittently within the cirque but becomes more prominent

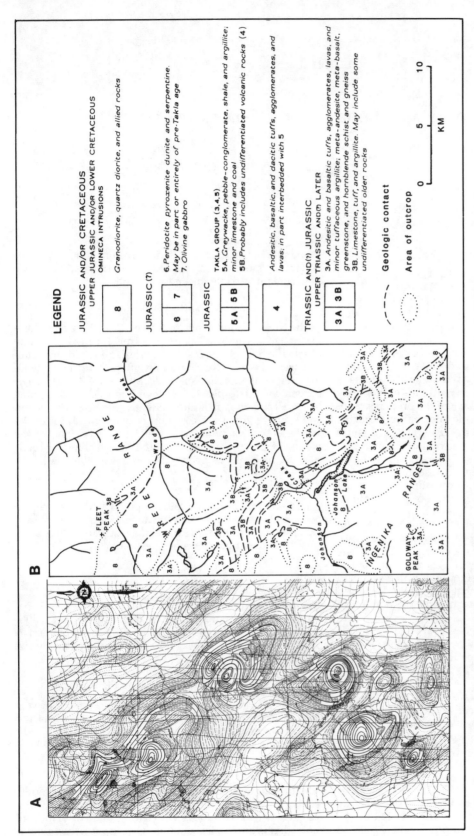

FIGURE 11.1.—(A) Aeromagnetic map (Geological Survey of Canada, 1972) and (B) geological map (Lord, 1948) on a topographic base (Government of Canada, 1952). Reproduced with permission of the Minister of Supply and Services, Canada.

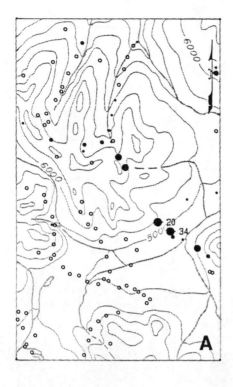
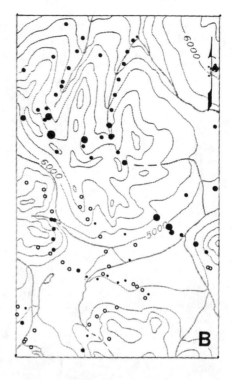
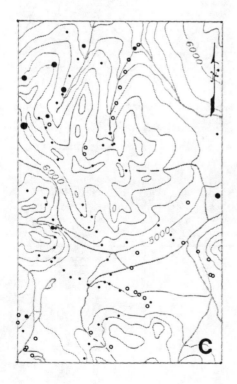

FIGURE 11.2—Results of regional stream sediment survey for (A) Mo, (B) Cu and (C) Zn. All values in ppm.

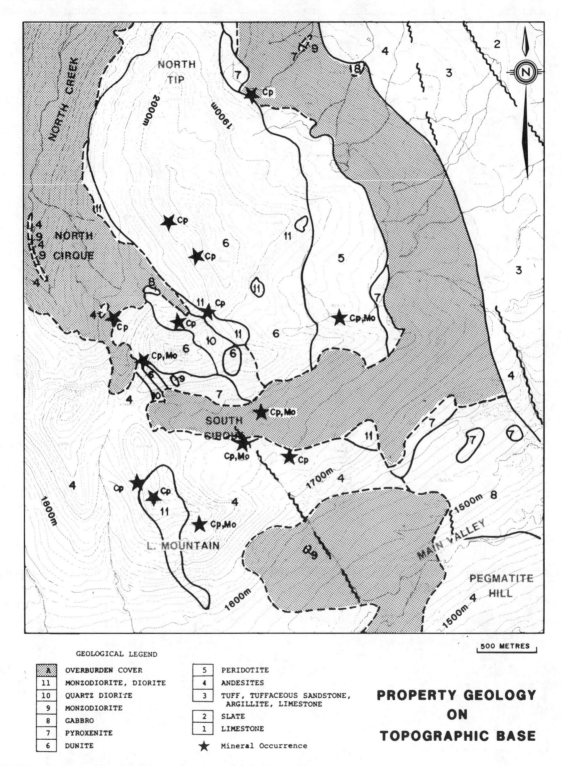

FIGURE 11.3—Geology and orthophotograph topography for the property.

towards its lip adjacent to the Main Valley. It is also relatively abundant along the main creeks: a prominent occurrence of molybdenite in bedrock was located along one of these creeks during followup.

Main Valley

The Main Valley is drained eastward by a large stream flowing within alluvial banks. Outcrop is exposed at the southwest and northeast extremities of the valley, and outcrop, covered by thin till, characterizes the base of Pegmatite Hill. Lateral moraines form a series of northeastward trending ridges along the north side of the valley, particularly below the lip of the South Cirque.

L. Mountain

Talus fans cover the mountainside. Bedrock is occasionally exposed despite a thin cover of till at lower elevations.

Pegmatite Hill

Topography and overburden type are similar to L. Mountain. Talus blocks contain chalcopyrite, bornite and molybdenite in pegmatite veins. Seepage zones are common at the base of the talus slopes.

Soils

Soil development over much of the property is weak or nonexistent above 1700 m. Below this elevation scrub spruce and lodgepole pine forest, with an understory of grass and widely spaced alder, have stabilized the surface and promoted formation of BM and BF horizons, in sandy soils, beneath a 2–3 cm thick LH zone. Strong Fe accumulations characterize BF horizons of soils at the foot of talus fans on Pegmatite Hill.

1. What are the principal overburden environments on the property? List at least five and describe what influences they might have on trace element dispersion.

2. What are the distinctly different geological environments to consider in interpreting geochemical results? How would each play a role?

SEMIREGIONAL EXPLORATION

In view of the relatively large area with mineral potential, a program of semiregional exploration, using stream sediments and reconnaissance soils (talus fines) near the base of the talus slopes, is a relatively obvious first choice from examination of property topography and air photographs. A more detailed stream and seepage survey was undertaken to better indicate locations of Cu and Mo anomalies. Almost 200 stream sediment and about 20 seepage samples were collected. Histograms showing the distributions of Mo, Zn, and Cu are illustrated in Figure 11.4. These were used to establish the size symbol intervals for Figures 11.5A–C. Note that seepage data have been plotted with the sediment results but are distinguished by symbol type.

3. Can you explain how the concentration levels were selected for the size coding of symbols? Do you agree with this?

4. If your answer to Question 3 is no, what contour levels would you recommend for each element?

5. Was it correct to use the same concentration values for stream sediment data as for seepage data. If not, what would be your principal concern? What would be your method of evaluating the two types of data?

CONTINUED OFFICE PLANNING

A government topographic map is available only at a scale of 1:250,000 and a contour interval of 150 m (500 feet) as shown in Figure 11.2. It is inadequate for controlling detailed work. An orthophotograph was therefore prepared at a cost of $3000 (Figure 11.6). The topographic map of Figure 11.3 was prepared from the orthophotograph.

Map scales were selected at 1:5000 for detailed work and 1:10,000 for compilation purposes. A baseline was planned to follow the apparent structural control to the mineral occurrences trending northwestward. Crosslines were established at a 150 m spacing and the sample interval was to be 50 m. The same lines were used to control magnetic and IP-resistivity surveys.

Three samplers were retained to collect approximately 100 soil samples per square km, with additional stream sediment or seepage samples to be taken whenever drainages were encountered along grid lines. About 2000 samples were to be collected at an average rate estimated at 40 samples/man day—requiring 50 man days to complete the survey.

6. Soil survey parameters for the property are summarized above. Do you agree with the choice of grid orientation, line spacing and sample interval? What might be some of your concerns in planning this survey?

7. What soil horizon would you sample? What soil horizon(s) would you avoid?

8. In 1973, only Mo, Cu, and Zn were determined on a routine basis. At a minimum, what elements would you have liked to have had data for in 1973, recognizing that (without ICP) each determination was likely to raise analytical costs by a minimum of $0.50 per element (Pb, Ag, Ni, Co, Fe, Mn, Cd) or $4.00 per element (Au, As, Sb, Bi, F, Cr, W, Sn). Soil pH determination cost $1.00. Why would you propose determining additional elements?

9. Based on the information available do you foresee any geochemical condition(s) likely to promote development of false anomalies?

PROPERTY EVALUATION

Property evaluation consisted of geological mapping, geophysical surveys and soil sampling.

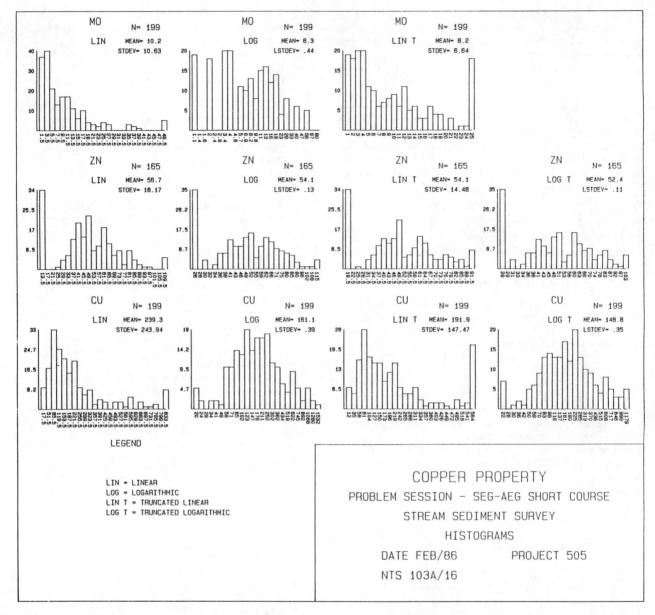

FIGURE 11.4—Histograms for Mo, Cu and Zn contents of stream sediments. The uppermost 5% of the data were omitted to calculate histogram intervals for the truncated (LIN T and LOG T) histograms.

Geology

Many areas of geochemical interest are concealed by overburden (Figure 11.3). Mapping suggests that most of the property is underlain by ultramafics with a central core of dunite surrounded by peridotite and pyroxenite and with hornblendite/gabbro to the south (Figure 11.3). A major structure trends northwestward through the middle of the North and South Cirques. West of this structure, bedrock is predominantly andesitic tuffs and breccias. Similar volcanic units underlie the western, northeastern and southern parts of the property.

Occurrences of chalcopyrite or molybdenite are found in granodiorite and quartz diorite boulders in the North and South Cirques, apparently related to suboutcropping dikes and plugs of these rocks. Cu grades of 1–2%, accompanied by argillic and advanced argillic alteration, make the search for their source an attractive target. Similar but lower grade Cu occurrences are found in dikes cutting ultramafic units on the North Tip.

Geophysical Surveys

Magnetometer and induced polarization surveys were initially considered to be the key to unravelling geology in overburden covered portions of the property. Ultramafic

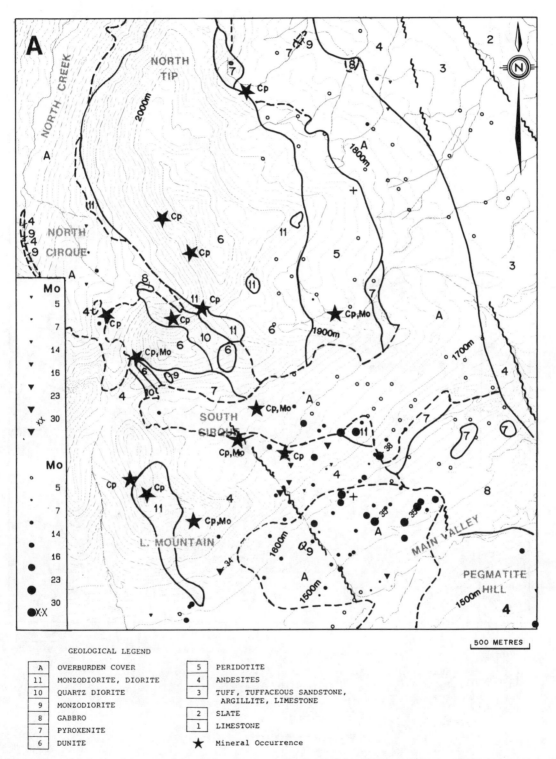

FIGURE 11.5—Stream (circles) and seepage (triangles) sediment results for (A) Mo, (B) Cu and (C) Zn. All results in ppm.

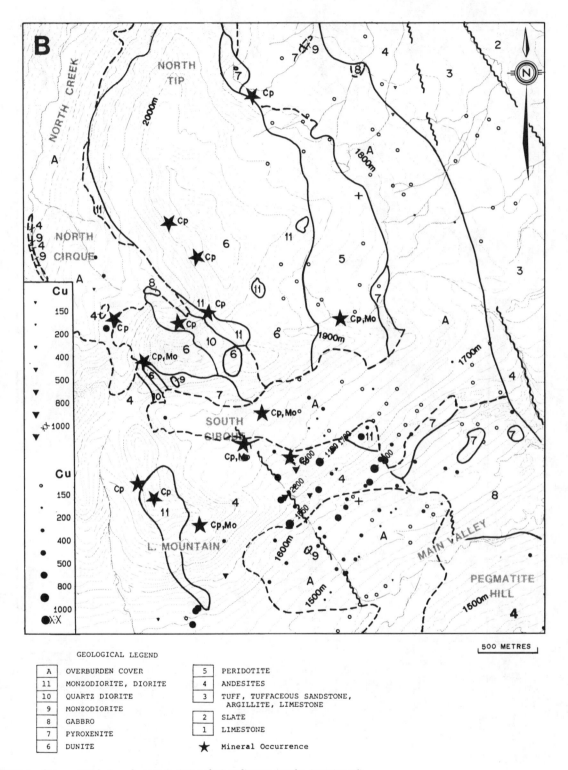

FIGURE 11.5—Stream (circles) and seepage (triangles) sediment results (continued)

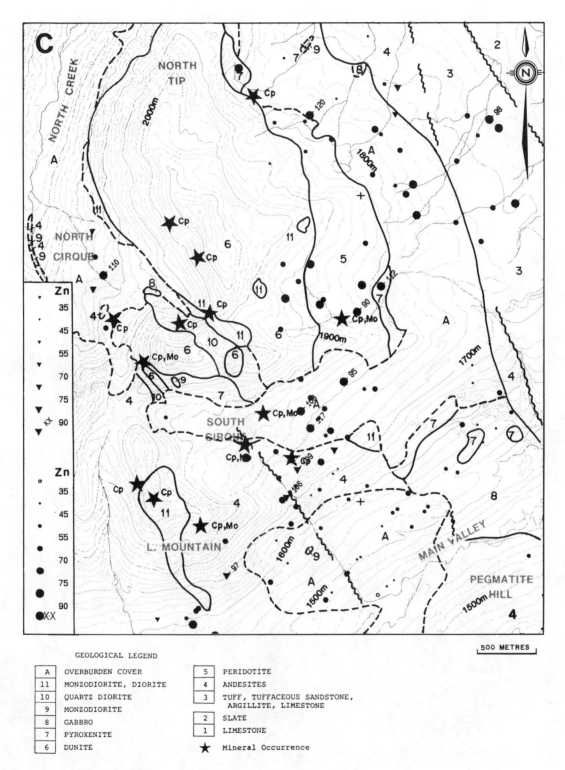

FIGURE 11.5—Stream (circles) and seepage (triangles) sediment results (continued)

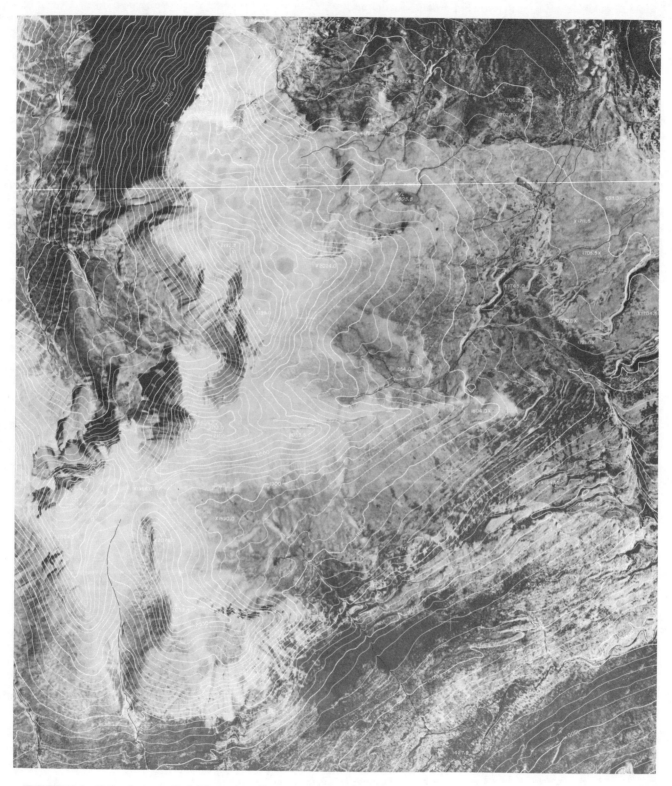

FIGURE 11.6—Orthophotograph of the property. Contours are superimposed in white.

rocks should be easily differentiated from volcanic and intrusive lithologies. However, when several geophysical anomalies at the head of the North Cirque were drill tested to obtain ground truth it was found that: (1) the dunite had a low magnetic signature, and variability within the ultramafic units was too great over narrow intervals to enable accurate prediction of underlying rock type, and (2) IP surveys gave very prominent anomalies, with strong responses from barren dunite and serpentinized ultramafic units that masked the signatures of sulfides in volcanic or intrusive rocks. Resistivity measurements were also unable to provide a focus for further work.

Soil Geochemistry

The landscape was such that both normal soils and talus fines would have to be collected. Concern that these two media would have different background levels led to evaluation of histograms for each separately (Figures 11.7 and 11.8, respectively). These were interpreted to provide size coding intervals for Mo, Cu and Zn data shown in Figures 11.9A–C, respectively, using circles for soils and diamonds for talus fines.

10. Are the defined size symbol levels satisfactory? Was it necessary to subdivide data on

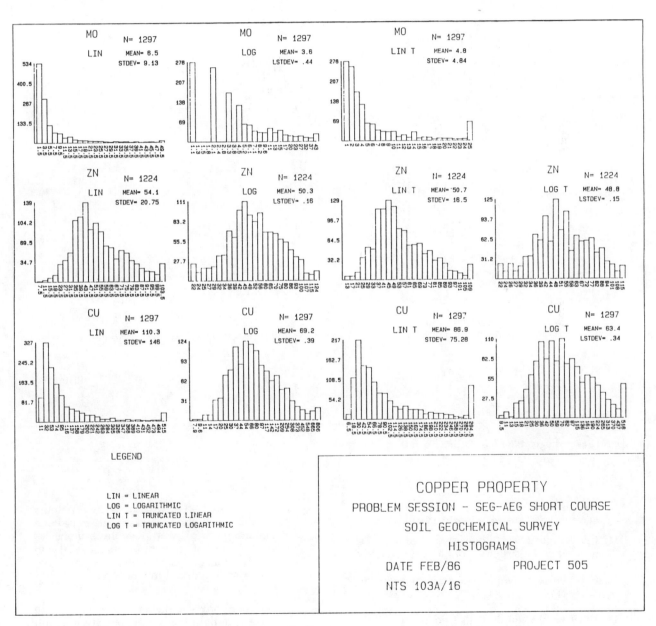

FIGURE 11.7—Histograms for Mo, Cu and Zn contents of soils.

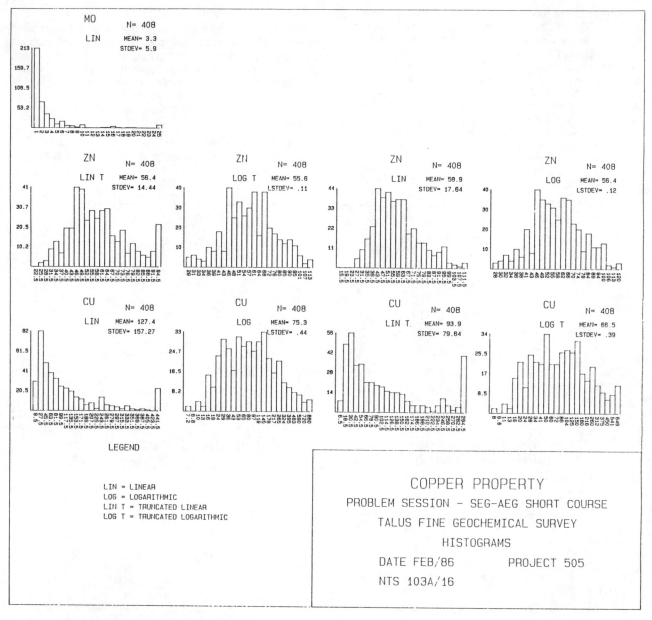

FIGURE 11.8—Histograms for Mo, Cu and Zn in talus fines.

the basis of the two types of soils? If you disagree, suggest the procedure you would choose.

11. Can any geologic units be interpreted from the geochemical distribution of Mo, Cu or Zn?

12. Can you identify any analytical errors in the data set?

13. What would you do with 1 point anomalies? What significance do you think they have?

14. On Figures 11.9A–C use one or two contours to define geochemical anomalies considered significant for followup. Rate these anomalies in the context of other information available to you, including logistical considerations. What factors are important in your decision(s)?

15. Are there any comments you would like to make regarding the sample density in South Cirque? Was it necessary?

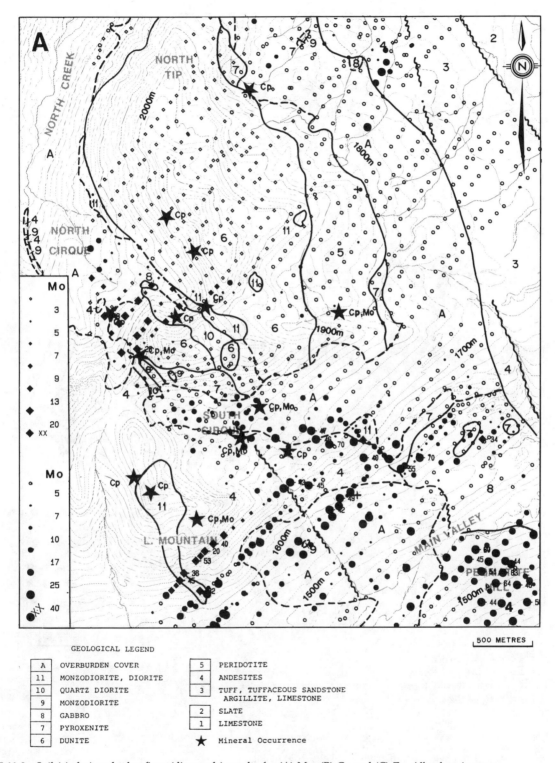

FIGURE 11.9—Soil (circles) and talus fines (diamonds) results for (A) Mo, (B) Cu and (C) Zn. All values in ppm.

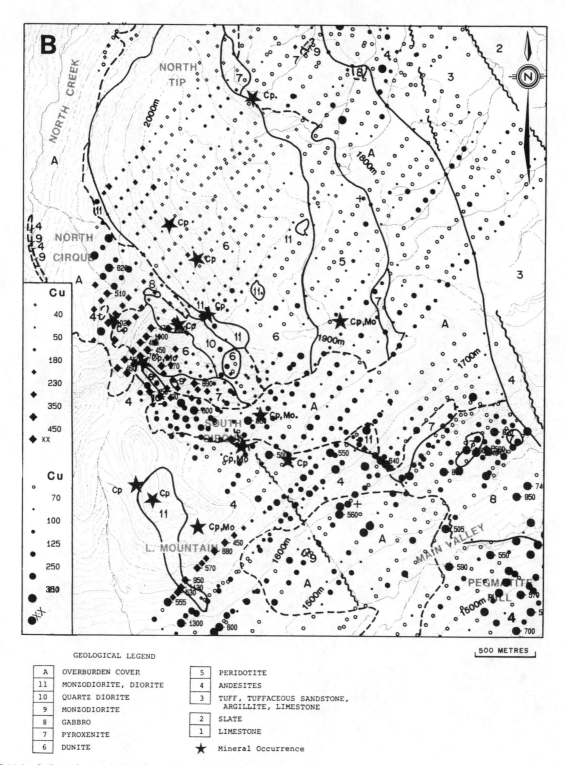

FIGURE 11.9—Soil (circles) and talus fines (diamonds) results (continued)

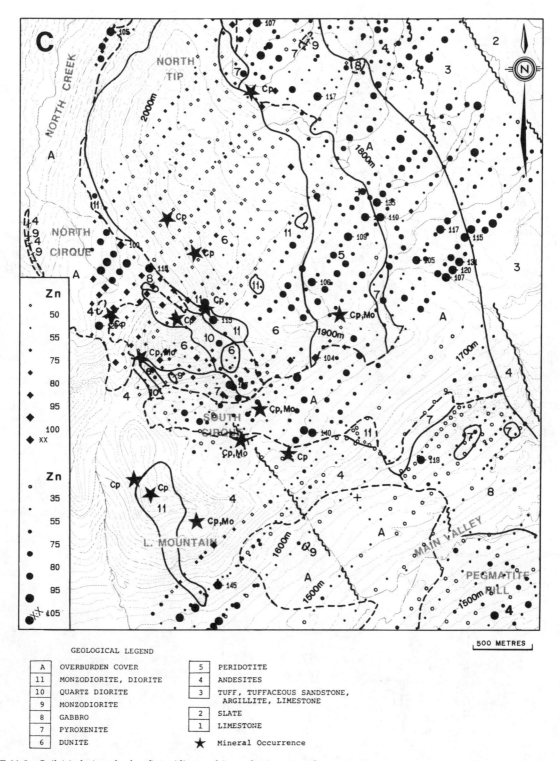

FIGURE 11.9—Soil (circles) and talus fines (diamonds) results (continued)

ANOMALY FOLLOWUP—DRILL TESTING

At this stage, the property had been mapped, prospected, geophysically surveyed and soil sampled. One of the five diamond drill holes sited to test geophysical anomalies in the North Cirque intersected 30 m of 0.4% Cu in weakly altered granodiorite below 50 m of pyroxenite. This was considered encouraging in view of its remoteness from the known Cu/Mo bedrock prospects and higher grade of boulder trains in both North and South Cirques. A model based on geological and geophysical information was proposed to guide drill testing (Figure 11.10).

Before additional drill testing can begin geochemical anomalies and mineralized boulder trains have to be rated and the probable locations of bedrock sources of Cu and Mo predicted. This may require additional followup work prior to drilling.

16. Use arrows to indicate probable dispersion pathways for metals (labelled G for glacial, L for landslide and H for hydromorphic dispersion) from possible bedrock sources to the zones defined in Question 14. Use R for residual anomalies.

17. Propose followup methods to precede a diamond drill program to evaluate areas outlined in Question 16. Rate your proposals as inexpensive, moderately expensive or expensive.

18. Design a followup program and prepare a budget to examine the anomalies assuming total funds are $350,000. A road extends to within 1 km of the southern boundary of the property. The following are approximate costs for different activities:

    (1) Grid preparation $300/man day

    (2) Soil and drainage survey
         collection costs $30/sample

    (3) Geophysical survey $750/line km

    (4) Geological field work $300/man day

    (5) Chemical analysis (including
         sample preparation) $15/sample

    (6) Road building/trenching $6,000/km

    (7) Diamond drilling, helicopter
         assisted $150/m

    (8) Diamond drilling, no helicopter $120/m

    (9) Supervision and report
         preparation   10% of total

    (10) Office overhead   8% of grand total

Your proposed exploration survey should be superimposed on your summary map of Question 14. Note any information you lacked and indicate if these omissions will have a significant impact on results.

ANSWERS

1. Recognition of major overburden environments suggests possible subdivisions of the geochemical data. Subsequent interpretation will have to determine if these significantly affect definition of anomalous conditions. In this case the following environments are recognized:
   a. Outcrop ridges—residual geochemical models.
   b. Talus slopes—mechanical transport downslope; seepage zones at base of slopes.
   c. Landslide areas—mechanical transport

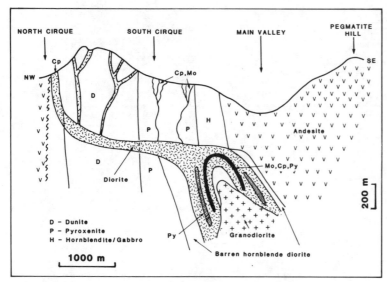

FIGURE 11.10—A schematic geological model for the property.

downslope, mixing of talus with valley floor deposits (till).
d. Glacial till—local glacial dispersion trains to be expected northwestward in the North Cirque, southeastward in South Cirque and more extensive transport east or west along Main Valley.
e. Alluvium—along the creeks draining Main Valley and North Creek. Alluvium should mask geochemical responses from underlying bedrock, except perhaps in areas of significant topography where hydromorphic dispersion has led to accumulation of metal in seepage zones.
f. Residual soils—over the Tabletop Highlands and North Tip; residual geochemical model.

2. Major lithological units that might be expected to influence geochemical responses are:
a. Ultramafics: subdivision is possible into areas underlain by dunite, pyroxenite, peridotite and hornblendite/gabbro.
b. Volcanics in the west, south and northeast.
c. Intrusive lithologies on the North Tip.

Each geologic unit is likely to contain a different background of trace, minor and major elements. In view of the locally derived character of much of the overburden on the property, metal distributions in soils should principally reflect metal contents of bedrock units. Relationships might, however, be somewhat modified by the influence of alkaline conditions in overburden derived from ultramafics versus weakly acidic conditions associated with weathering sulfides in acid intrusives. Contrasting rock types, containing very different backgrounds of elements such as Ni, should enable estimates to be made of the distance and direction of overburden transport and possibly the extent of mixing of overburden derived from different sources.

3. The objective of contouring or coded symbols is to focus attention on areas of high mineral potential.

Only a relatively small proportion of the survey area should be highlighted as anomalous. Contours are first chosen to separate distinctive populations on histograms or probability plots (Chapter 5). The upper 5% and 10% tails of each population are then usually highlighted—though choice of the percentiles is entirely subjective. A total of six size coded symbols optimizes symbol size differences, firstly to be readily distinguishable to the unaided eye and secondly to be small enough to avoid cluttering on maps. The lowest values are represented by an open symbol instead of a dot to ensure stray marks introduced by map reproduction are not mistaken for geochemical results. In addition, the highest numbers of the survey, whether they are significant or not (i.e., samples taken on top of known but trivial mineral occurrences; poor samples producing false anomalies, etc.), are emphasized using the largest symbol with the concentration value annotated. This avoids the otherwise usual question "How high is the highest value?".

4. The above guidelines were followed and therefore a major change would not be suggested.

5. It is not appropriate to use the same contour levels. Inspection of Figure 11.5A & B indicates that levels of Cu and Mo in seepages are generally significantly higher than values in nearby streams. This reflects the different geochemical controls in the two media. To consider them the same for interpretive purposes is wrong. Always separate obviously different materials into recognizable classes and evaluate their geochemistry independently.

6. Sulfide occurrences are most abundant in the North and South Cirques, approximately following the contact separating volcanic and ultramafic lithologies. A grid baseline should follow this orientation. The geologic target is probably some type of bulk tonnage Cu–Mo deposit that, to be economically attractive, must lie relatively close to surface. In the landscape environment provided by the two cirques, it would have to have a relatively large surface expression beneath the overburden. A sample interval of 100 to 150 m would be appropriate to outline an anomalous area, closer spacing representing overkill at this stage.

7. The answer to this question would be dependent on the outcome of an orientation study. However, the BF or BM soil horizons might be anticipated to give the best anomaly contrast in vegetated areas. The C1 horizon, representing unaltered overburden, would be taken where soil profiles are not developed. Organic matter accumulations in boggy areas would be avoided as would samples having unusual and/or non-representative colors or textures.

8. Because of its high background content in ultramafics, Ni should be included to provide information on the distribution of ultramafic derived overburden and on its mixing with volcanic derived overburden in the cirques. This could be helpful in estimating effect of dilution on soil Cu values—assuming a strictly mechanical mixing. Cr determinations might perform the same function, but analytical costs are higher and results possibly less reliable than for Ni.

Routine determination of soil pH would also be desirable. Ultramafic units are associated with soil pH's of 8 to 9.5 which would prevent hydromorphic dispersion of Cu. However, alkaline conditions promote Mo mobility. Weak Mo anomalies in an alkaline environment, in nonseepage areas, might therefore be more significant than Mo anomalies in acidic environments.

Control exerted by Fe on the generation of false Mo and perhaps Cu anomalies can be evaluated if Fe data are available. The information would be particularly important in seepage zones in forested areas. Soil color determination might suffice as an alternative to Fe determination.

The additional analyses suggested would have cost about $2.00 per sample or an increase of 66% over the $3.00 for the Mo, Cu and Zn determination.

9. From the limited information available the following possibilities might be suggested:
   a. Scavenging of Mo and perhaps Cu by iron oxides or organic matter in seepage areas.
   b. pH barriers, accumulation of Cu if the metal was leached under acidic conditions and migrated in groundwater into an alkaline surface environment. Mo would accumulate in the transition zone from alkaline to acidic conditions.

10. Contour levels are considered satisfactory for this problem. The data could be reexamined in selected areas, based on your responses to Questions 1 and 2, to maximize identification of weak anomalies in each environment. For Mo and Cu (but not Zn) it was necessary to subdivide the data into soil and talus fine groups for interpretation.

11. The method of determining concentration levels for size coding of symbols highlights geological controls. Anomalies within each geological domain are reflected by locally larger symbols, although not necessarily by the largest symbols for the survey. In this case, the following geological and geochemical relationships are apparent in Figures 11.9A–C:
   a. Dunite—low Cu (<50 ppm), low Zn (<50 ppm) except on its western margin where the dunite is enriched in Cu (50–180 ppm).
   b. Pyroxenite—moderately high Cu (150–250 ppm).
   c. Pyroxenite/peridotite—high Zn (80–120 ppm).

These patterns are further emphasized by contouring of multisample features in Figure 11.11.

12. Analytical errors are frequently recognized by systematic variations along a single line. Problems might be suspected south of L. Mountain, but further inspection shows that the line of very high values correlates with talus debris at the base of slope and is therefore probably environmentally controlled. Soils immediately downslope were sampled in a different overburden environment (till) beyond the influence of the talus deposits. Obvious analytical errors are not apparent in the data.

13. Once orientation studies have defined optimum sample densities, one point anomalies should have little significance on a detailed soil grid, even if a metal value is exceptionally high. This is particularly true in this case because the target is a large bulk tonnage deposit. Isolated high values may be due to poor sampling (false anomalies) or proximity to an insignificant mineralized boulder or bedrock occurrence. Reference should be made to soil coding forms to identify sampling-introduced problems and then prevent their recurrence on future surveys. Bona fide anomalies in the case history under review are thus multisample features.

14. Geochemical anomalies worthy of followup may be defined by examining Figure 11.9 and noting multisample features exceeding local background. These have been contoured and shaded on Figures 11.11.A–C. Figure 11.12 was then compiled by combining Cu and Mo anomalies. It is apparent that a majority of the Cu–Mo anomalies lie south or west of the ultramafic intrusion. Geochemical patterns for Zn show a concentric zoning around the dunite core of the ultramafic intrusion and do not appear to be directly related to known chalcopyrite-molybdenite occurrences. Elsewhere, Zn data have insufficient local contrast to assist in rating Cu–Mo anomalies.

Bedrock exposure is inadequate to map the geology of anomalous areas, and geophysical surveys have been unable to resolve the complexities of the geology. Thus a working approach to target definition is based on the coincidence of abundant mineralized boulders and outstanding Cu and/or Mo soil anomalies tempered by considerations of local access. In order of priority six anomalous areas, numbered (1) through (6) in Figure 11.13, can be outlined:

(1) North Cirque*: Abundant high grade boulders; high Cu and Mo in soils.
(2) South Cirque*: Moderately abundant high grade boulders; Mo bedrock prospect; high Cu, moderate Mo in soils.

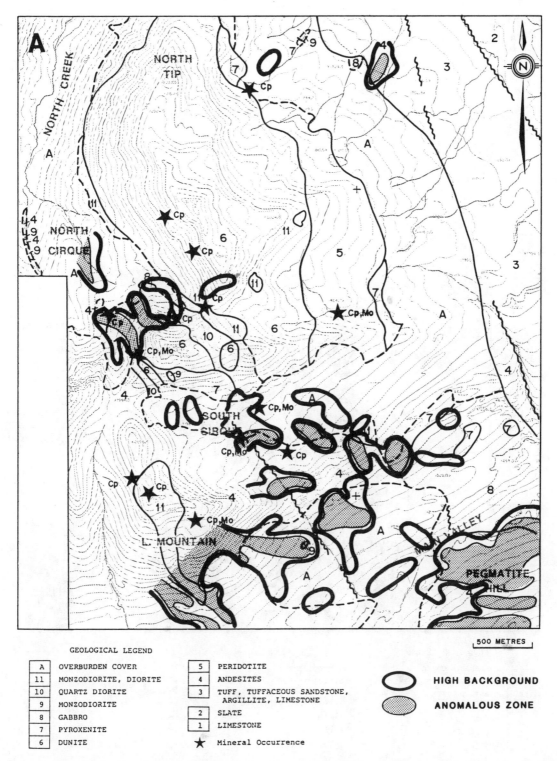

FIGURE 11.11—Contoured geochemical maps summarizing distributions of high background and anomalous conditions for (A) Mo and (B) Cu. High background areas only shown in (C) for Zn.

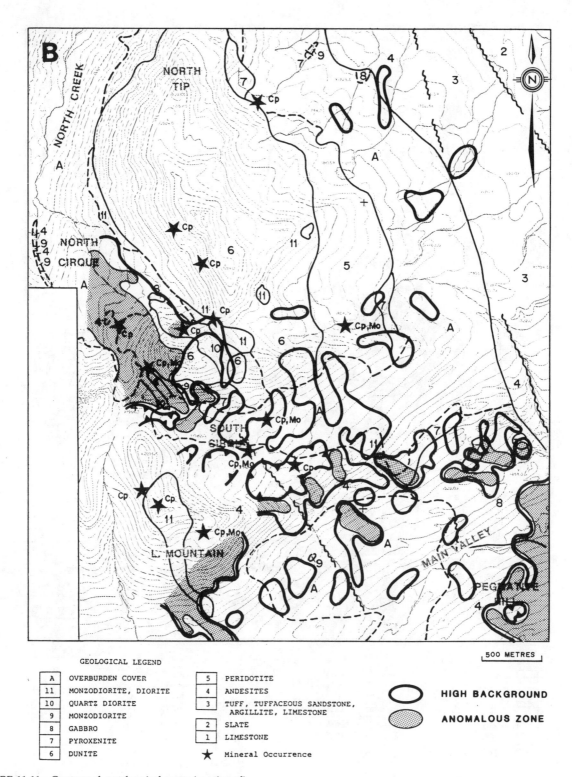

FIGURE 11.11—Contoured geochemical maps (continued)

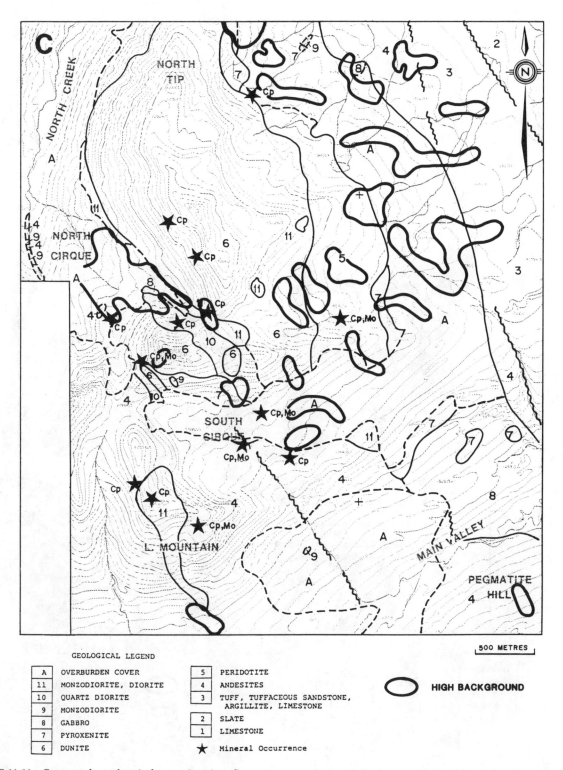

FIGURE 11.11—Contoured geochemical maps (continued)

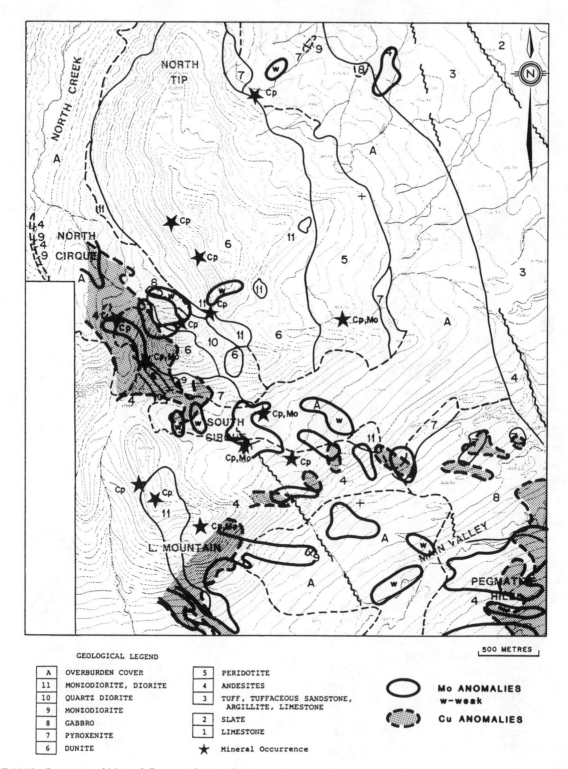

FIGURE 11.12—Summary of Mo and Cu anomalies on the property.

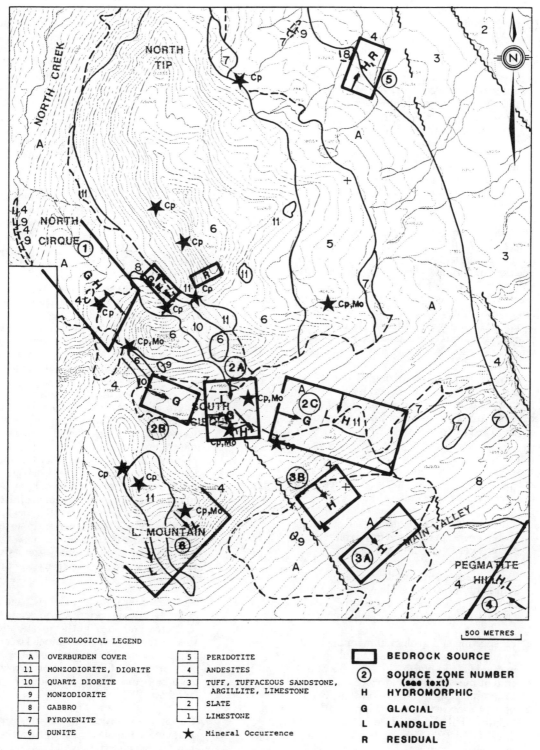

FIGURE 11.13—Proposed bedrock source areas for Mo and Cu soil anomalies. Suggested dispersion mechanisms from the bedrock sources to the anomaly sites are indicated.

(3) Main Valley: Heterogeneously high Cu and/or high Mo in an alluvium filled valley.
(4) Pegmatite Hill: Chalcopyrite-molybdenite bearing pegmatites nearby; high Mo, moderate Cu in soils.
(5) North Tip: Linear Mo soil anomaly, weak Cu; mineralized dikes nearby.
(6) L. Mountain: High Cu and Mo, weakly mineralized boulders in talus fans. Grades are probably representative of bedrock upslope.

*Priorities for North and South Cirque would be reversed on consideration of easier access, thinner overburden and absence of solifluction features in South Cirque.

15. The South Cirque was sampled at twice the density of North Cirque. The additional sampling was unnecessary in view of the composition and origin of the overburden and the anticipated size of the surface expression of the sought-after target. The extra samples do not materially change definition of anomalous conditions. Money expended on the extra sampling and analysis could have been spent elsewhere.

16. See Figure 11.13, summarizing potential bedrock source areas for the six anomalies defined in Question 14. The dispersion mechanisms believed to have transferred Cu and Mo to the soils are indicated.

17. Many possible followup programs could be recommended. The following is a series of representative procedures to be considered or avoided in your selection.

   a. Additional, more detailed soil sampling would be inexpensive but unwarranted: see the answer to Question 15.

   b. Deep overburden sampling to explore for root zones of surface soil anomalies would be moderately expensive and very difficult because of the abundance of talus boulders.

   c. Prospecting and geological mapping might be effective in evaluating L. Mountain, Pegmatite Hill and North Tip. Absence of outcrop at the bottom of North Cirque, South Cirque and Main Valley will limit usefulness of prospecting and mapping at these locations but should nevertheless be conducted. Distribution and abundance of mineralized boulders (and different styles of mineralization, if any) could provide drill targets in view of the local origin of the overburden.

   d. Trenching to bedrock might be appropriate in the South Cirque where overburden is thin along the Main Valley and north of Pegmatite Hill. The program would be moderately expensive, considering the need for an access road, but cost benefits might accrue by using a bulldozer to simultaneously prepare access roads for a possible drill program. Considerable logistical difficulties would be encountered in attempting to trench North Cirque, and a program of diamond drilling could probably be completed at lower cost. Trenching of the one anomaly at North Tip would probably also be excessively expensive.

   e. Reanalysis of existing soil sample pulps using the same digestion procedure (i.e., aqua regia) but a multielement determination method (e.g., ICP) would be inexpensive and likely to aid further interpretation. Reanalysis for gold today might locate gold anomalies that would significantly upgrade economic prospects for the property. Gold geochemistry might aid recognition of residual geochemical features and distinguish them from hydromorphic patterns. Reanalysis would probably only have to consider costs for about 1000 samples at an average cost of $10 to $15 per sample.

   f. Soil pH, if not already known, could be determined relatively inexpensively and would establish the environmental conditions for large areas on the property. This information should aid interpretation of the mobility of elements such as Cu and Mo and suggest areas where spurious accumulations are occurring at pH barriers. It might also allow for interpretation of hydromorphic or mechanical genesis of an anomaly, which could then be investigated further using partial extractions. Correct interpretation of anomaly genesis will optimize ground prospecting and/or trenching.

   g. Geochemical sampling of all seepage zones in the North and South Cirques and north of Pegmatite Hill might focus attention on specific localities within a much larger area. The procedure is inexpensive but will only be worthwhile if seepages are relatively abundant and evenly distributed on the landscape.

   h. Geophysical surveys. Detailed ground magnetics and IP-resistivity surveys have not proven effective. A VLF survey would be inexpensive, but interpretation of results would have to consider the influence of topography. Resistivity surveys to deter-

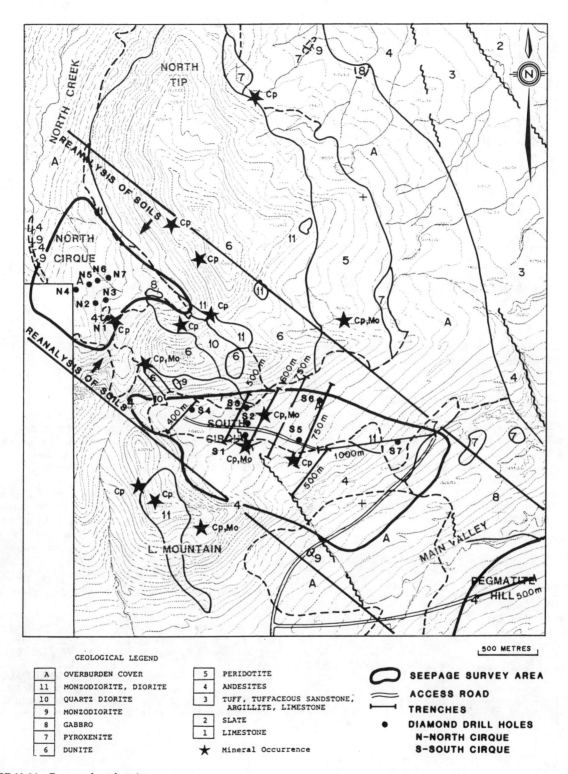

FIGURE 11.14—Proposed exploration program.

mine overburden depths and identify near surface zones of bedrock alteration might be appropriate. Proposals to use more expensive ground electromagnetic surveys could be considered if objectives are clearly defined.

18. Many possible programs could be suggested for the property. In the one that follows (Figure 11.14) no further soil sampling is recommended because it is believed that the property is at the diamond drill stage and that the majority of the funds must be expended on this activity. Preliminary studies using soil geochemistry or routine geophysical surveys are over: it would still be possible to extend the grid to the west, but if this was merited it should already have been accomplished. Discovery of more anomalies on new ground will only confuse the issue—sufficient interest has already been generated with available data to get on with the property evaluation. A possible work program is indicated in the following budget summary:

Preliminary Studies

A. Reanalysis of existing soils by a multielement procedure +Au +pH:
   1000 samples @ $15/sample $15,000

B. Seepage survey, North and South Cirques and Pegmatite Hill:
   300 samples @ $ 30/sample collection $ 9,000
   300 samples @ $15/sample analysis $ 4,500

C. Anomaly ground evaluation (geology, prospecting), North Cirque (3 days), South Cirque (3 days), Main Valley (2 days), Pegmatite Hill (2 days), North Tip (1 day), L.Mountain (1 day)
   Total 12 days @ $300/day $ 3,600
   100 samples @ $15/sample analysis $ 1,500

   SUBTOTAL: $33,600

Physical Work

A. Access road:
   estimate 6 km @ $6,000/km $36,000

B. Trenching (Main Valley, South Cirque and Pegmatite Hill)
   5 km @ $6,000/km $30,000

C. Mapping and sampling trenches
   20 days @ $300/day $ 6,000

D. Bedrock chip analysis from trenches
   500 samples @ $15/sample $ 7,500

   SUBTOTAL: $79,500

Diamond Drilling

A. South Cirque
   7 holes @ 100 m each
   700 m @ $120/m $70,000
   400 samples @ $15/sample $ 6,000

B. North Cirque
   7 holes @ 100 m each
   700 m @ $150/m $105,000
   400 samples @ $15/sample $ 6,000

   SUBTOTAL: $187,000

Project Management, Report Writing
   10% of $299,600 $30,000

Office Overhead
   8% of $329,000 $26,000

   TOTAL: $356,100

SUMMARY

This case history illustrates the many interrelated variables that must be considered during interpretation of soil surveys. Recommendations must ensure that followup funds are well spent examining bona fide anomalies. The likely bedrock source for an anomaly must be predicted, and it is a serious error to assume that contoured high values are a "bullseye" for the bedrock source of metals. Failure to correctly identify anomaly sources at an early stage can seriously distract the exploration effort resulting in lost time and money: at worst the project may fail.

REFERENCES

Geological Survey of Canada. 1972. McConnel Creek, British Columbia. Aeromagnetic Map Sheet 94D.

Government of Canada. 1952. McConnel Creek, British Columbia. Map Sheet 94D, National Topographic Series, Third Edition.

Levinson, A.A. 1980. Introduction to Exploration Geochemistry, Second Edition. Applied Publishing Ltd., Wilmette, Illinois, 924 pp.

Lord, C.G. 1948. McConnel Creek Map Area, Cassiar District, British Columbia. Geological Survey of Canada Memoir 251, 72 pp.

Rose, A.W, Hawkes, H.E. and Webb, J.S. 1980. Geochemistry in Mineral Exploration, Second Edition. Academic Press, 657 pp.